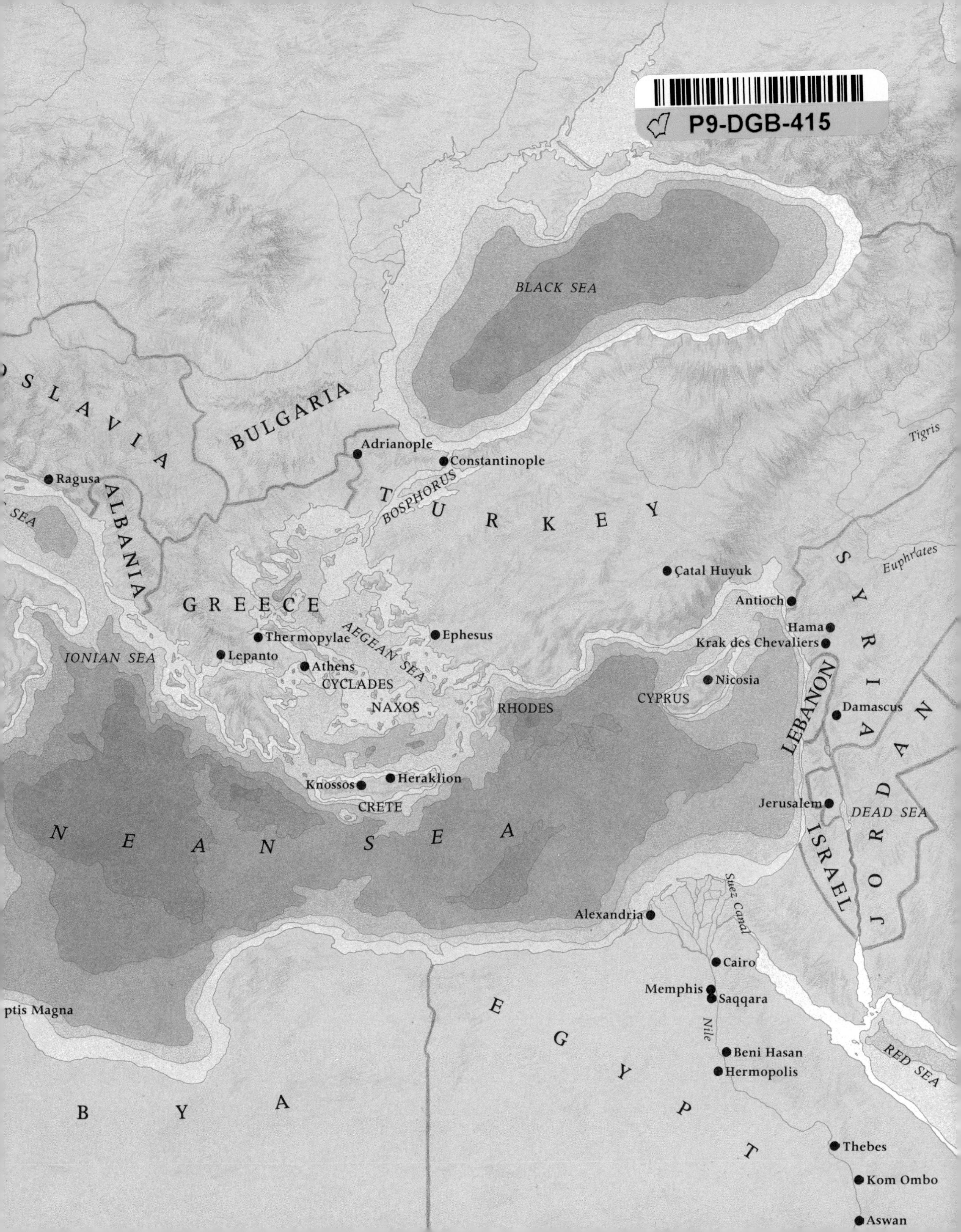
BLACK SEA
BULGARIA
ALBANIA
Ragusa
Adrianople
Constantinople
BOSPHORUS
TURKEY
Tigris
Euphrates
SYRIA
GREECE
AEGEAN SEA
IONIAN SEA
Thermopylae
Lepanto
Athens
CYCLADES
NAXOS
Ephesus
RHODES
Çatal Huyuk
Antioch
Hama
Krak des Chevaliers
Nicosia
CYPRUS
LEBANON
Damascus
JORDAN
Jerusalem
DEAD SEA
ISRAEL
Knossos
Heraklion
CRETE
Suez Canal
Alexandria
Cairo
Memphis
Saqqara
Nile
Beni Hasan
Hermopolis
EGYPT
RED SEA
Thebes
Kom Ombo
Aswan

THE FIRST EDEN

THE FIRST EDEN

THE MEDITERRANEAN WORLD AND MAN

DAVID
ATTENBOROUGH

LITTLE, BROWN AND COMPANY
BOSTON TORONTO

First American Edition

Published by arrangement
with William Collins Sons & Co. Ltd.

Printed in the United States of America
RRD OH

Barbary Macaques in the cedar forest of Morocco

CONTENTS

Foreword – 7

PART ONE

THE MAKING OF THE GARDEN

The filling of the sea – 10   The population of the waters – 19
The animals of the islands – 29   The changing climate – 33
Surviving the summer – 39   The arrival of mankind – 58

PART TWO

THE GODS ENSLAVED

The sacred animals of prehistory – 64   The animal gods of Egypt – 72
The kingdoms of the bull god – 95   The butchery of the gods – 111

PART THREE

THE WASTES OF WAR

The taming of the horse – 122   The triumphs of Islam – 132
The animal myths of Europe – 140   The migrations of the Crusaders – 148
Eastern animals move west – 160   The destruction of the forests – 166

PART FOUR

STRANGERS IN THE GARDEN

Marine invaders from the east – 177   Terrestrial invaders from the west – 182
The human invasion – 194   The plunder of the wildlife – 198
The relics of Eden – 208

Acknowledgements – 232

Picture Credits – 233

Bibliography – 234

Index – 236

FOREWORD

Eden, some will argue, must have had a precise geographical location, and it was probably a small, sandy island somewhere in the Red Sea. But the Garden of Eden in our minds and our myths is surely the primeval wilderness where human beings first wandered; where, in the words of the Bible, grew every tree that is pleasant to the sight and good for food. There Adam gave names to all living creatures. There one of his sons learned how to domesticate sheep, and another started to till the ground and grow crops for the first time. If that is so, then there is good justification for regarding the whole of the Mediterranean world as humanity's first Eden, as I have done in giving this book its title.

My original intention, when I first started thinking about the book and the television programmes connected with it, was to survey the natural history of this fascinating and beautiful part of the world, where so many of us go for our holidays, and where we have a more sustained opportunity to wander in wild country and look at animals and plants than we do in the places where we work and live. It could, therefore, have become solely a description of the glamour spots of European natural history, the reserves where the fauna and flora of the Mediterranean still survive in very much their pre-industrial condition. But to have restricted a survey to that would have been to describe only a tiny proportion of the Mediterranean countryside and to ignore the bulk of the places where most of us spend most of our visits. Those landscapes can only be understood in the light of the past. This, after all, is the oldest humanised landscape in the world. Nowhere has mankind had a greater effect on his environment or left more continuous, detailed and abundant evidence of his activities. So this book began to include more and more archaeology and history, and became an attempt to describe how our attitude to the natural world, and our treatment of it, has changed over the centuries and has produced the living communities we see today.

Such an examination could, of course, be made of any tract of land on earth; but the Mediterranean has a special claim to our interest. It is not merely a popular holiday resort. It is the place where mankind's exploitation of the land began, and where it has run its full cycle. What happened here during past millennia is, elsewhere on earth, just beginning.

PART ONE

THE MAKING OF THE GARDEN

Native trees growing on ancient vine terraces in Greece.

THE FILLING OF THE SEA

Six and a half million years ago, the Mediterranean Sea vanished. The arm of the ocean that, for tens of millions of years previously, had separated Europe from Africa suddenly dried out, exposing an immense, irregular trench in the earth's crust, two thousand miles long and maybe as much as a mile deep. On its hot, airless floor lay a string of shallow briny lakes, rimmed with white deposits of salt. A few volcanic mountains, that today form Corsica, Sardinia and some of the Greek Cyclades, rose sufficiently high above the valley floor to attract rain, They, therefore, were clothed with forests in which lived a variety of animals. But for the most part this gigantic furrow in the earth's surface was roastingly hot, and almost lifeless.

The evidence to support this astounding vision lies today beneath the sparkling blue waters of the Mediterranean itself. It was hauled up from the depths in 1970 by scientists working on the American research ship, the *Glomar Challenger*, who were investigating the structure of the sea-floor. Wherever they drilled, often in as much as 1,000 feet of water, they encountered, at around five hundred feet down in the rocks of the sea-bed, a layer of salts so thick that it continued downwards beyond the reach of their drilling equipment. To measure its full extent, they had to use echo-sounding and other sophisticated techniques. It proved, in places, to be over a mile thick.

The samples brought up by the *Glomar Challenger*'s drills were, chemically, somewhat mixed. As well as ordinary salt, sodium chloride, which mineralogically is known as halite, they contained calcium sulphate or gypsum, and anhydrite, a form of calcium sulphate that crystallises without incorporating molecules of water. Minerals such as these are called, as a group, evaporites, for they occur wherever shallow salty water is evaporating in a strong sun – in the southern end of the Dead Sea, for example, and in the marine lagoons that fringe the Persian Gulf. Crystals of halite and gypsum are among the first to appear as the water gets saltier and warmer. Anhydrite, the most soluble of all, only crystallises when the water temperature rises above 35° Centigrade, almost as warm as human blood. Its presence, therefore, is a certain and eloquent indicator of what conditions were like when it formed.

The distribution of these different salt minerals in the rocks of the Mediterranean sea-floor convincingly supported the hypothesis that the sea had once been reduced to a group of shrinking salt lakes. In the shallower parts of the Sea – the first areas to be exposed as the water evaporated – the salt is predominantly anhydrite. Halite and other soluble salts, such as potash and magnesium salts, which only crystallise when the waters are at their briniest, are largely restricted to the deepest parts of the basin, where the last remnants of the evaporating salt lakes would have lain.

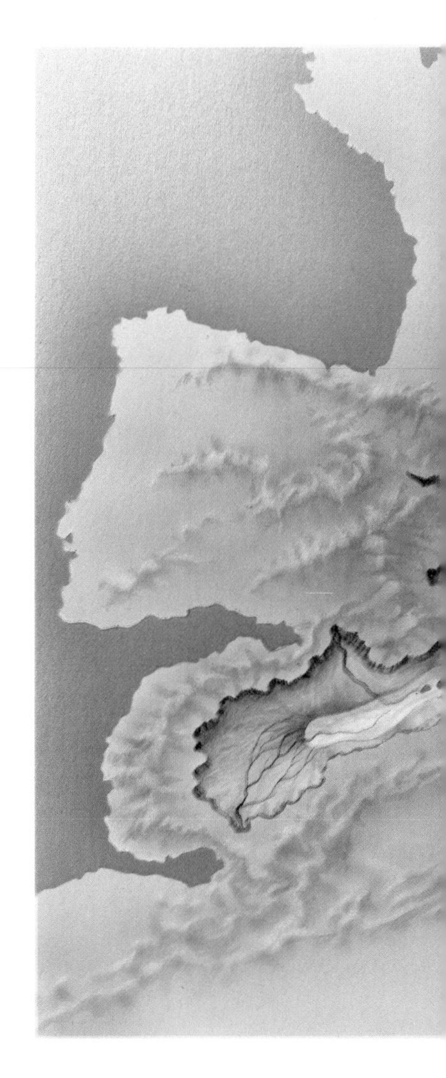

An evaporated sea
When the Mediterranean was at its driest, its waters were reduced to a few briny lakes in its lowest parts, as shown in this artist's reconstruction (*below*). The Dead Sea (*left*) acquires the salts which crystallise in its shallows and coat boulders around its shores, from subterranean sources and not from the evaporation of sea water. Even so, it gives a good idea of how the Mediterranean salt lakes must have looked five million years ago.

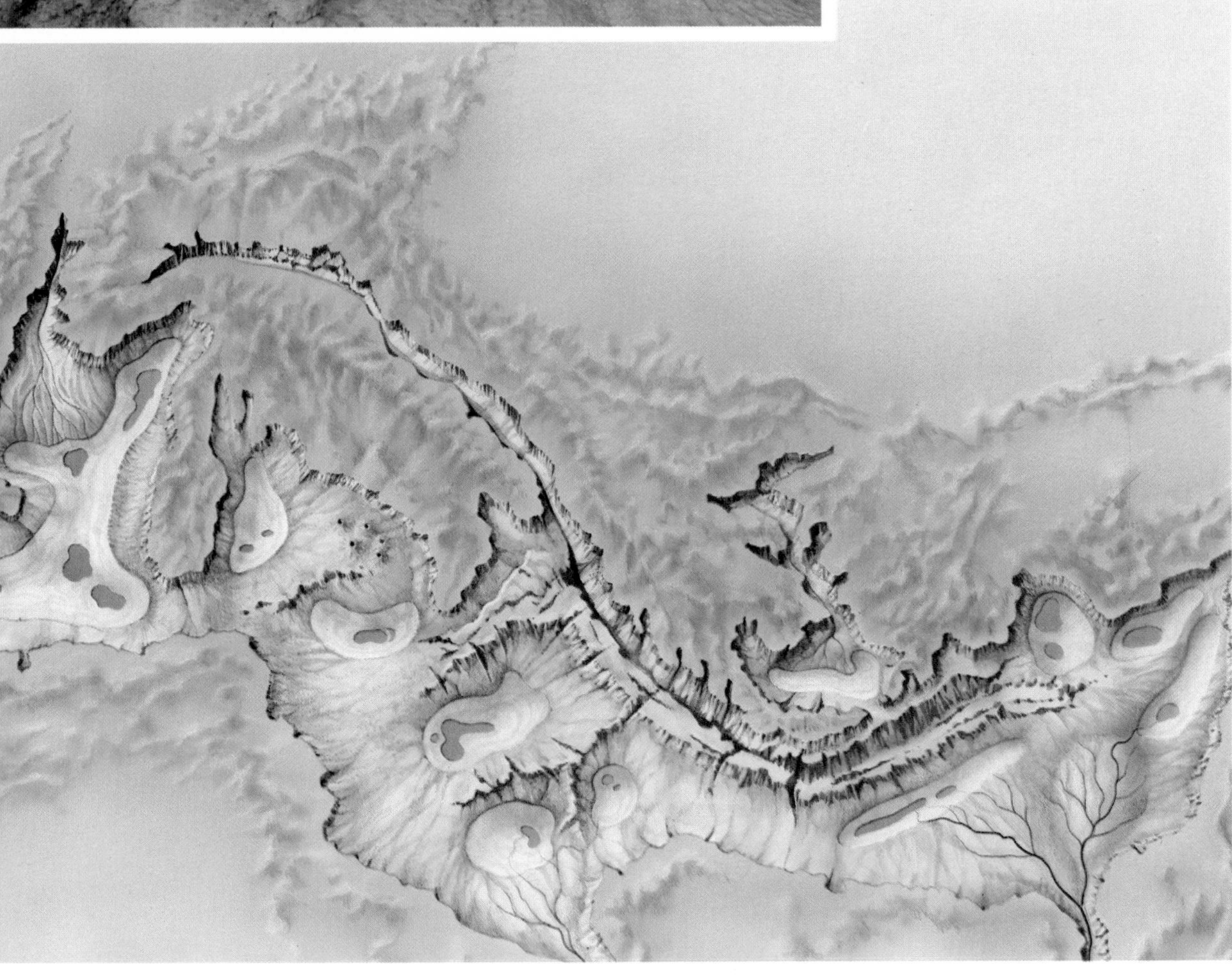

The possibility that these salt deposits could have formed in some strange way in deep water was ruled out by evidence within the cores themselves. Among the salts, the scientists found the microscopic remains of algae. Algae are plants, and like all green plants they depend upon light for their existence. They cannot live in deep water. So there could be no doubt – shallow lakes had once existed on the floor of the Mediterranean.

Further corroboration that the sea had once vanished came from geological facts that had long been known but which, until now, had seemed inexplicable. The river Rhône in southern France flows through a plain of sands and gravels, but buried far beneath these deposits, cut into the granite bed-rock, lies a deep channel. It was first discovered by engineers who were drilling for water. It can be traced down the Rhône valley, sinking ever deeper as it goes, until near the Mediterranean coast it is 3,000 feet below the present sea-level. A similar buried channel underlies the valley of the Nile. Five hundred miles upstream at Aswan, it can be detected 700 feet below the present land surface. In the delta, it sinks so far beneath the thick layers of sediments deposited by the river that it seemed untraceable. But recently, oil geologists located it. It lies 8,000 feet beneath the city of Cairo. For decades, scientists puzzled how such gorges, cut far below the present level of the Mediterranean Sea could have been produced by rivers. Now we know. After the sea had disappeared and the Mediterranean basin was dry, the Rhône and the Nile tumbled down steep cliffs onto the floor of the oven-hot basin a mile or so below and in doing so gouged out these deep courses for themselves.

But how could the Mediterranean have lost its waters?

The continents are not static on the earth's surface. They are composed of rocks that are lighter in weight than the black basalt which forms the ocean floor and which underlies the continents themselves. Basalt, though a heavy, hard and unyielding rock in the hand, nonetheless slowly deforms when subjected to great pressure and heat over long periods. Indeed, viewed over the vast perspective of geological time, it can be regarded as a very viscous liquid on which the continents float like rafts. Heat from the earth's immensely hot core is thought to create currents that rise and circulate in the basalt and, very very slowly, drag the continents over the surface of the globe.

Sixty-seven million years ago, Africa, with Arabia still attached to its north-eastern flank, was an island and widely separated from Europe. But as the continents moved, so that gap slowly closed. The sediments that had been eroded from the surface of the land by the rivers and spread by them over the sea-floor between the approaching continents, began to crumple and ruck up, millimetre by millimetre,

The sea-floor raised
As the continental blocks of Africa and Europe approached one another, the muds and sands that lay on the bottom between them were compressed and crumpled to form mountains like these in southern France, in which the once horizontal layers are still clearly visible.

century after century. Some, compacted by the great pressure into rock, emerged above the surface of the waters. But still an extensive, if relatively narrow sea lay between the two continents.

Forty million years ago, the sediments had been raised so high they formed a range of mountains, the Alps, and the collision was slowing down. The basic shape of the basin that would eventually become the Mediterranean had now been determined, even though it was still connected with the rest of the oceans of the world, both at its eastern and western ends.

Some twenty million years ago, Africa started to collide with the lands of the Middle East, and the straits at the eastern end began to close. Some five million years later, the fish and other organisms living in the waters between Africa and Europe were cut off from those in the Indian Ocean, and from this time onwards these two marine communities evolved in isolation from one another. The animals and plants living on the land on either side of the new junction, however, now met for the first time and began to mingle. Antelope and horses that had evolved in Eurasia moved south into Africa. Monkeys and elephants, which were of African origin, invaded Europe and Asia.

But the movement of the continents had not ceased. The north-western coast of Africa now came into contact with the southern tip of Spain. The Mediterranean was sealed at both ends. As long as it remained so, it was doomed to desiccation.

Even in today's relatively cool climate, a thousand cubic miles of water evaporates from the surface of the Mediterranean each year. If the Gibraltar Straits were dammed now, the Sea would dry up in around a thousand years, even if the Rhône and the Nile continued to flow into it. Six million years ago, the climate was somewhat hotter than it is today, so the waters must have disappeared even more swiftly. The isolation, however, was not unbroken. On repeated occasions, the waters of the Atlantic spilled over the neck of land that connected Morocco and Spain. Several pieces of evidence show that this happened. For one thing, the huge thickness of salt deposits discovered by the *Glomar Challenger* to underlie almost the entire Mediterranean is far greater than could have been produced by the evaporation of a single basin-full of Mediterranean water. Some calculations suggest that forty times that amount of water would have been needed to produce so much salt. More evidence comes from the salt deposits themselves. They are not uniform throughout their thickness, but made up of many layers, separated by bands of muddy material. This must indicate several partial refillings of the basin after periods of evaporation. In one major and widespread discontinuity, the salt immediately beneath one such muddy layer has an eroded

upper surface, indicating that here it had been exposed to the air and that winds carrying gravel and sand had swirled across its surface, turning it to powder and blowing it away, before another flood arrived, carrying more mud from the surrounding hills. Judging from sections bored through the whole deposit, there seem to have been several of these cycles of flooding, salt deposition and desiccation.

The removal of an immense weight of water from the Mediterranean basin must have caused major movements in this part of the earth's crust. The underlying quasi-liquid basalt layer is likely to have risen slightly, elevating the rocks of what was once the sea floor. Cracks developed along the edges of the continents and caused long series of earthquakes in the process. Old volcanoes became active again and new ones developed along fresh lines of weakness. About 5.3 million years ago, a particularly deep series of faults running east and west broke the compacted junction between Morocco and

A Mediterranean volcano
The first exploding mountain to carry the name Vulcano was a small islet a few miles north of Sicily which in Greek and Roman times regularly erupted with loud detonations and which was believed to be the forge of Vulcan, the blacksmith to the gods. Today it is quiescent; but a little farther north, Stromboli (*above and left*) 'the lighthouse of the Mediterranean' remains more or less continuously active, as it has been for centuries.

Gibraltar. The land shifted and fell. Once again – and for the last time – the Atlantic flooded across the isthmus connecting Morocco and Spain.

How high the cliffs forming the western wall of the Mediterranean basin were at this time is still under debate. Some geologists believe that they plunged steeply down for as much as 10,000 feet to the valley floor – about as far as from the summit of Mont Blanc to the surface of the Lake of Geneva. They would have been about fifty times higher than Niagara and probably extended horizontally for many miles. Water surged down these stupendous falls in a vast cataract at a rate, it has been estimated, of over 40 cubic miles a day. In about a century, a mere moment of geological time, the whole basin was filled once more with water. Few seas, or indeed any large-scale feature on the face of the earth, can have had such a dramatic, swift or precise a birth as today's Mediterranean Sea.

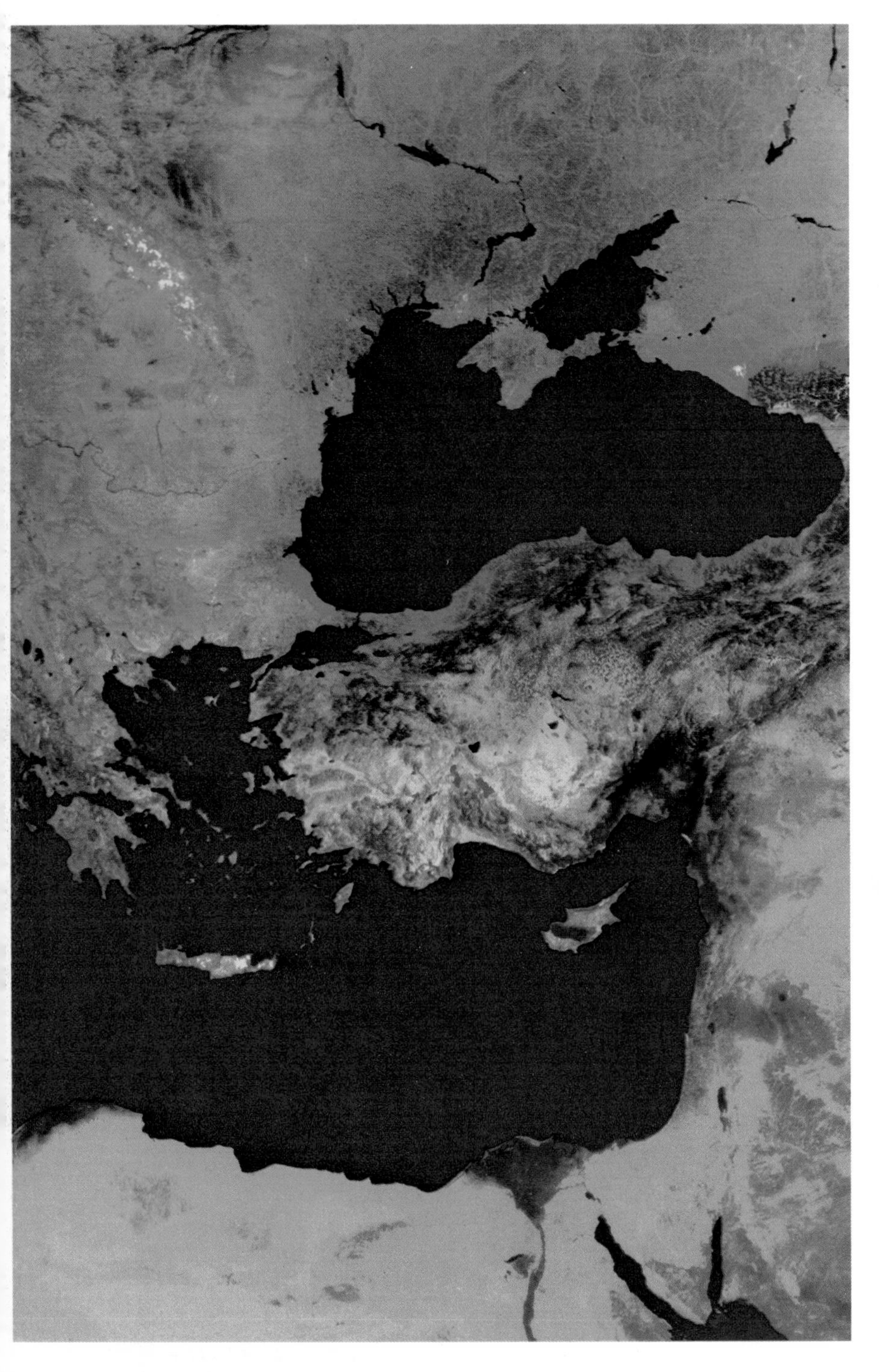

The Sea from Space
Several photographs taken by a weather satellite on almost cloudless days between 1979 and 1985 have been put together like a mosaic to produce this remarkable image. The output from three different sensors, each working on a different wavelength, has been synthesised to produce an approximation of natural colour.

3250 feet
6500 feet
10,000 feet
surface currents
sub-surface currents

THE POPULATION OF THE WATERS

The sea that thus came into existence was, and remains, an odd one. The Gibraltar Straits, at their narrowest, are a mere nine miles across and at one point no deeper than a thousand feet. This restricted connection with the Atlantic Ocean has given the Mediterranean a special character. As its waters evaporate under the warm sun, their saltiness increases. To compensate for this loss, water from the Atlantic streams in through the Straits at a speed of about 2.5 miles an hour. The saltier water, however, is slightly denser and it sinks downwards. Eventually, it flows out into the Atlantic. So in the Straits, two currents are moving simultaneously in opposite directions, one near the surface flowing eastwards into the Sea, and one at depth, flowing westward out of it.

The narrowness of the Straits also hinders the waters of the Sea from following those of the ocean outside as they rise and fall with the tug of the moon. Although the Mediterranean itself does not escape this pull, the rise it produces is not nearly so great as it would be if there were free access to the open oceans. This, together with the relative steepness with which the sea bed slopes away from most of its coasts, keeps Mediterranean tides relatively small.

These physical peculiarities of the Sea have affected the character of the animals and plants that live in it. As far as we know, no marine organisms survived the great desiccation. Those that occupied the newly-filled basin must, therefore, have all come from the Atlantic. The fact that, initially, they must have been swept over a gigantic waterfall and that any subsequent invaders had to cross a relatively shallow sill, has prevented many of the inhabitants of the really deep parts of the Atlantic from colonising the Mediterranean, even though there are areas of it that might well suit them. Off the west coast of Italy there is a basin 12,000 feet deep, and in the deepest part of the whole Sea, south of Greece and Turkey, a series of narrow trenches sink down to more than 16,000 feet. But these black waters contain far fewer species than are found at such depths in the ocean outside. Instead they are inhabited by molluscs, bristleworms and other organisms most of which appear to have invaded these underpopulated depths from the shallower parts of the sea.

No such obstacle, of course, faced the creatures from the shallower parts of the Atlantic and they must have colonised the newly-created Mediterranean very swiftly. Today in the Sea east of Gibraltar, just as in the Ocean to the west of it, red mullet plough through the sediments searching for food, and skate and plaice lie almost invisible on the sands; in rocky areas, lobsters and crabs clamber among the boulders and octopus lurk in holes; and hunting in the middle waters above swim squid and shark.

Close in to the shore, the comparatively small tides allow plants to

Gateway into the Sea
A satellite view of the Gibraltar Straits showing Spain to the left and Africa to the right, with the Atlantic in the foreground and the Mediterranean stretching away into the distance. The Rock of Gibraltar itself is the tip of the tiny promontory just inside the Straits. (*Below*) Surface water from the ocean is continually flowing into the Mediterranean to compensate for evaporation, but denser saltier water is also flowing out at depth.

grow that cannot survive the daily exposure and submersion endured by the wracks and other marine algae growing on Atlantic shores. One of these, named Posidonia after the sea-god of the ancient Greeks, produces the piles of brown ribbon-like leaves that are so characteristic of Mediterranean beaches and the waving underwater meadows so familiar to Mediterranean divers. Posidonia is not an alga, but one of the very few flowering plants that have managed to invade salt water. It belongs to the Potomageton family, most of whose members live in fresh-water ponds. It grows in the coastal shallows and develops fleshy underground stems which form a thick mat, anchoring the sediment deposited in the sea by rivers. Its flowers sprout from the top of the stems within the curled sheath of the uppermost leaf. The stamens discharge their pollen in the form of long threads which float away to fertilise other flowers.

As in all seas, the upper waters of the Mediterranean contain plankton, that floating assemblage of microscopic plants, fish fry, sea urchin and crab larvae, shrimps and other animals, that constitute the basic source of organic food in the sea. Many of these organisms are swept in from the Atlantic by the surface current flowing through the Gibraltar Straits. Plankton produces a steady rain of dead organisms which drifts slowly downwards to the bottom where it forms a rich nutritious ooze. In most seas, this is regularly stirred up by storms and so refertilises the upper waters. The Mediterranean, however, loses a considerable proportion of this nutriment, for the current of dense salty water flowing continuously out of the Sea at depth, carries much of it away into the Atlantic. So the plankton of the Mediterranean is not as rich as in many oceans, and the crystal clarity of its blue waters, that delights so many swimmers is, in fact, an indication of their relative infertility.

Nonetheless, they contain quite enough plankton to support large shoals of anchovies and sardines and other small surface-feeding fish. These in turn attract predatory fish such as mackerel and their giant relatives, the tuna and the swordfish. These last two species can reach a length of twelve feet and a weight of over a thousand pounds. Most are not permanent residents of the Mediterranean but nomads that wander widely through the oceans of the world in search of their prey. The tuna travel up into the North Sea and regularly cross the Atlantic from the Bay of Biscay to the New England coast. One tagged individual made this trans-Atlantic crossing in only 119 days which, even if the fish had travelled determinedly on a steady and accurate course,

The microscopic community of the Sea
Algae (*right*: magnified 130 times) drifting in the sunlit waters are food for larval crustaceans (*above right*: magnified 20 times).

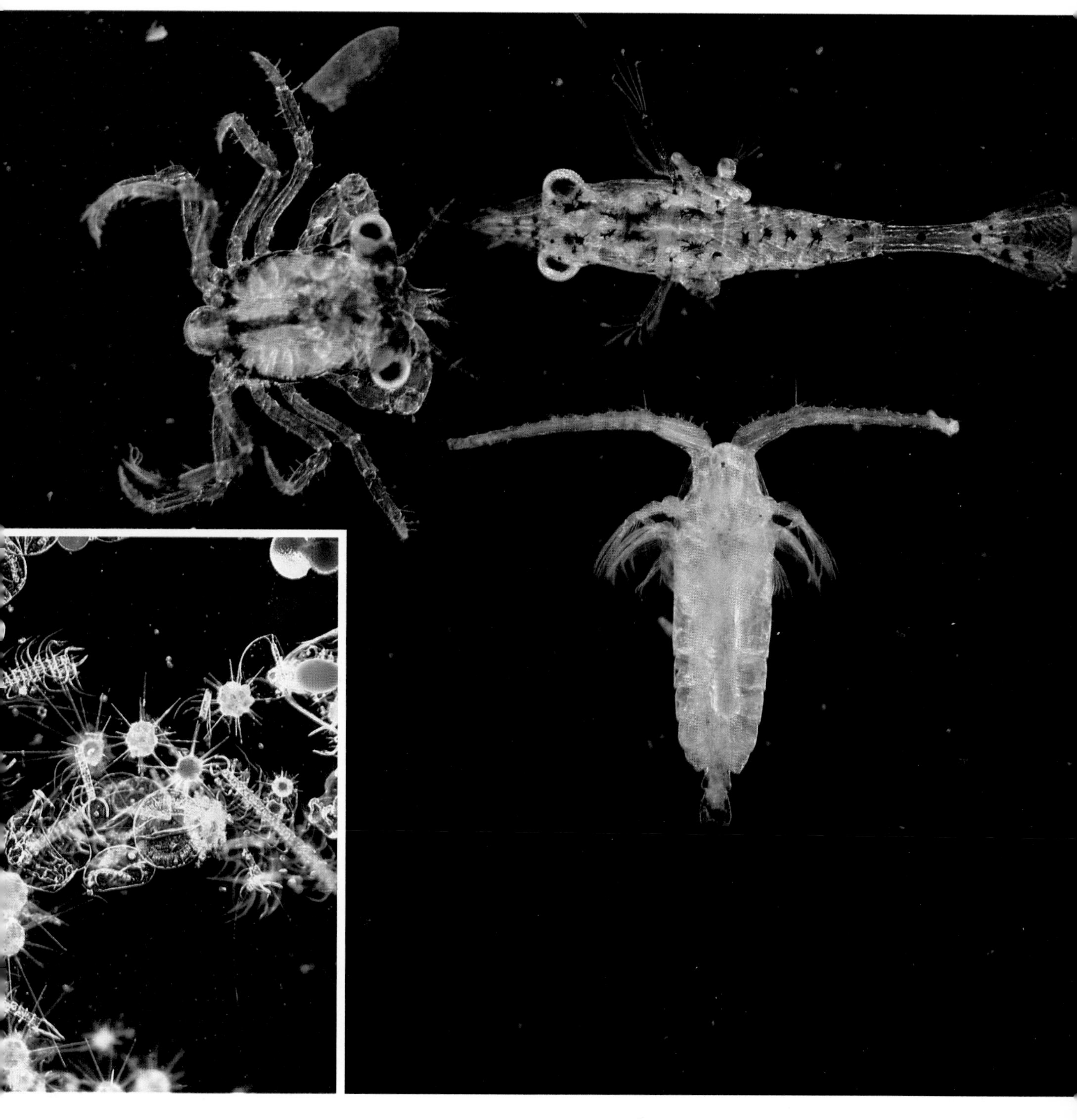

would have required it to cover at least 40 miles a day. Both of these superb and handsome fish come to the Mediterranean each year in order to spawn. The young remain in the Sea for at least the first year of their lives and probably more. Then they migrate westwards through the Gibraltar Straits to hunt in the Atlantic before returning in due course to spawn.

Other Atlantic animals also pay regular visits to the Mediterranean. The immense ocean-going leatherback turtle, which can grow to a length of six feet, whose global wanderings are still largely uncharted sometimes appears in the Sea, drifting through the water lazily browsing on shoals of jellyfish. Loggerhead turtles, which are very much smaller, come in to feed on the Posidonia meadows. Some haul themselves out of the water to lay their eggs in the sand of beaches around one or two of the Greek islands and along the remoter parts of the Turkish coast. A significant number have even become permanent residents and never leave the Sea at all.

Sea mammals come too. Pilot whales and killer whales are often seen passing through the Gibraltar Straits, but their invasion seems a somewhat tentative one, for they seldom travel farther east than Corsica. Sperm whales, with their immense flat-topped heads and narrow, toothed jaws, cruise in on regular migrations to collect squid and octopus. But the commonest sea mammals are the dolphin. Schools of them surge through the surface waters, leaping in the air in mass displays of what seems to be sheer *joie de vivre* as they chase the shoals of anchovies and sardines. Thousands assemble each summer near Gibraltar and stay there until autumn. No one knows why.

Seals, the most recent of the mammals to colonise the seas, also have their representatives here. They evolved, long before the Mediterranean dried, from otter-like ancestors, somewhere around the shores of the North Atlantic. One of the earliest groups to appear were the monk seals. They are the only seals that frequent warm sub-tropical seas. Some crossed the Atlantic, established a population in the Caribbean and went on, at a time when the Isthmus of Panama was submerged, to enter the Pacific and settle around the Hawaiian Islands. Others travelled less far and found their way into the Mediterranean where today they are among the rarest of the Sea's inhabitants. Their name 'monk' comes from a Greek word meaning 'solitary' for they are usually seen in ones or twos unlike other seals that congregate in herds.

So, in the five million years since its refilling, this youngest of seas has acquired a rich and varied population of inhabitants.

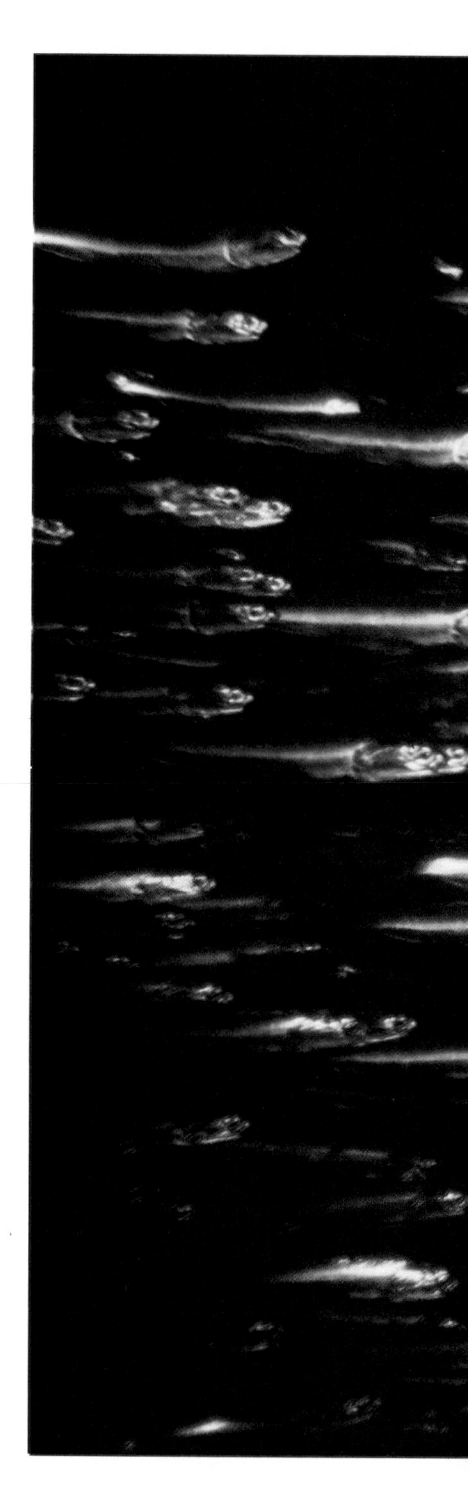

Feeders on plankton
Anchovies are only a few inches in length but they form vast shoals, millions-strong.

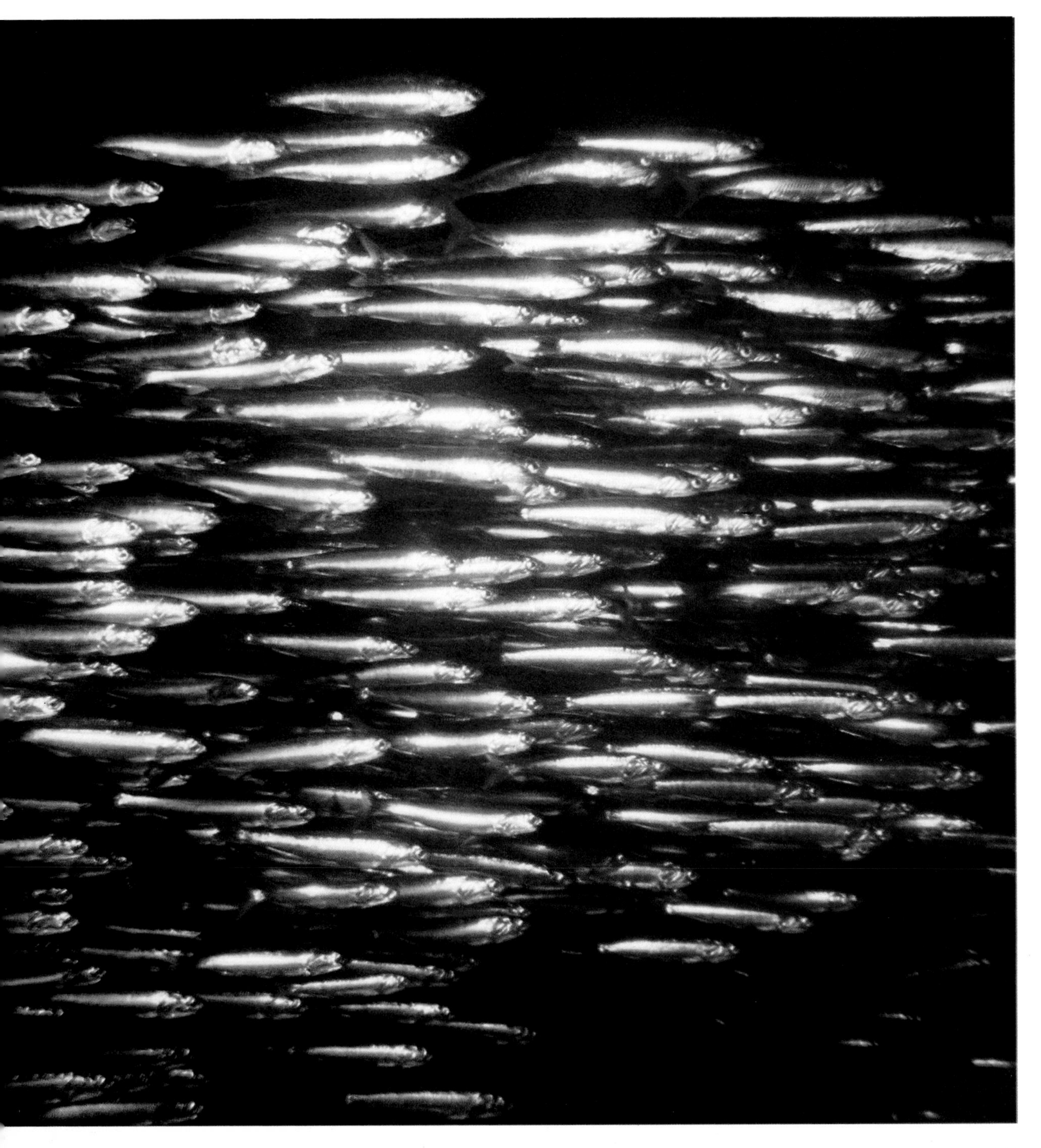

Fish for all levels

Flying fish (*above*) live in the upper waters. When they are pursued by predators, they shoot above the surface and can glide for several hundred yards on their outstretched pectoral fins. In middle depths cruise several species of shark including the blue shark (*far right*), which roam widely through the sea in search of their prey and may be attracted from long distances away by the presence of blood in the water. Other fish, like the grey mullet (*right*) seldom stray far from the bottom of the sea, where they grub about seeking invertebrates, marine algae and detritus.

Newcomers to the Sea
Several groups of air-breathing land animals have produced species that have colonised the sea. The turtles, such as the loggerhead (*below*), are reptiles. They are able to stay below the surface for long periods, but eventually they have to come to the surface to breathe. The monk seal (*below right*), the only seal species in the Mediterranean, is a mammal. It must come ashore to pup. Dolphins (*right*) are also mammals, but they have evolved a technique of giving birth at sea and helping their newborn to the surface to snatch the first breath.

THE ANIMALS OF THE ISLANDS

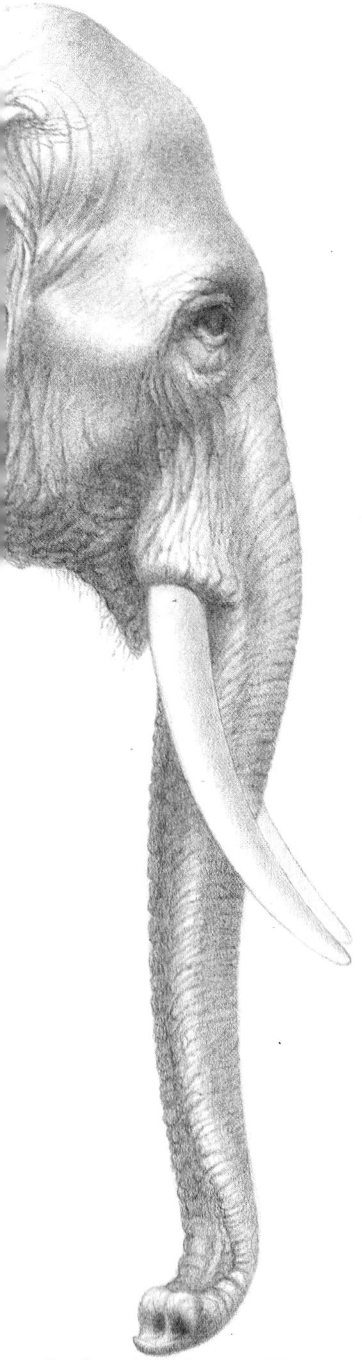

The mountains that stood above the empty Mediterranean were probably colonised by animals even before the sea returned. Their cool, green upper slopes must have provided a welcome refuge for any animal that strayed down from the continents on either side into the hot depths of the valley between. But even after the basin refilled and these mountains had become islands, animals continued to reach them. Some creatures, when swimming around the coasts of the continents may have been carried out to sea by the currents. Others, perhaps, were involuntary passengers on rafts of vegetation that floated down the Nile, the Rhône and other big rivers, and were eventually washed up on the beaches of the new islands. Certain it is, however, that soon after the refilling, all the islands of any size were inhabited by a range of animals that included between them mice and shrews, hedgehogs and deer, tortoises and land birds, and even hippopotamus and elephants. None, however, had any large carnivores such as lion or leopard, and the their absence was to become important.

Imprisoned on their islands, and faced with conditions crucially different from those their ancestors had experienced on the mainland, these animals began to evolve in their own particular way, just as animals on the Galápagos and other remote islands round the world have done. One of the best known of these animal communities is that which lived on the island of Malta. At a time when the rainfall was considerably greater than it is today, streams flowing over the limestone that forms the island began to trickle down the cracks, dissolving away the rock and eventually creating a long winding cavern. Today it is known as Ghar Dalam. The river which ran through it carried not only mud and pebbles, but also the remains of the animals that lived on the land above. Most of the bodies were dismembered and disarticulated by the rushing waters. A high proportion of the bones were smashed into unrecognisable fragments. But bigger bones such as shoulder-blades and limb bones, stronger ones such as knee-caps, and especially teeth, survived, packed into the gulleys and crevices, and give some idea of how remarkable this community of animals had become.

Three species of elephant lived on Malta. All were dwarfs. The biggest of them was only seven feet tall, the smallest a mere three feet, smaller even than a Shetland pony. They appear to be more closely related to the Asian elephant than to the African species. The reason for their small size may well be connected with the absence of predators on the island. Elephants in Africa and Asia rely for their immunity to attacks from lions and tigers on their huge size and tough skins. When they are no longer in such danger, their great bulk is no longer necessary for their survival and, on an island where there are limited supplies of vegetation on which to feed, large stature is difficult to reach and to maintain.

The dwarf elephant and hippo of Malta beside a modern Indian elephant.

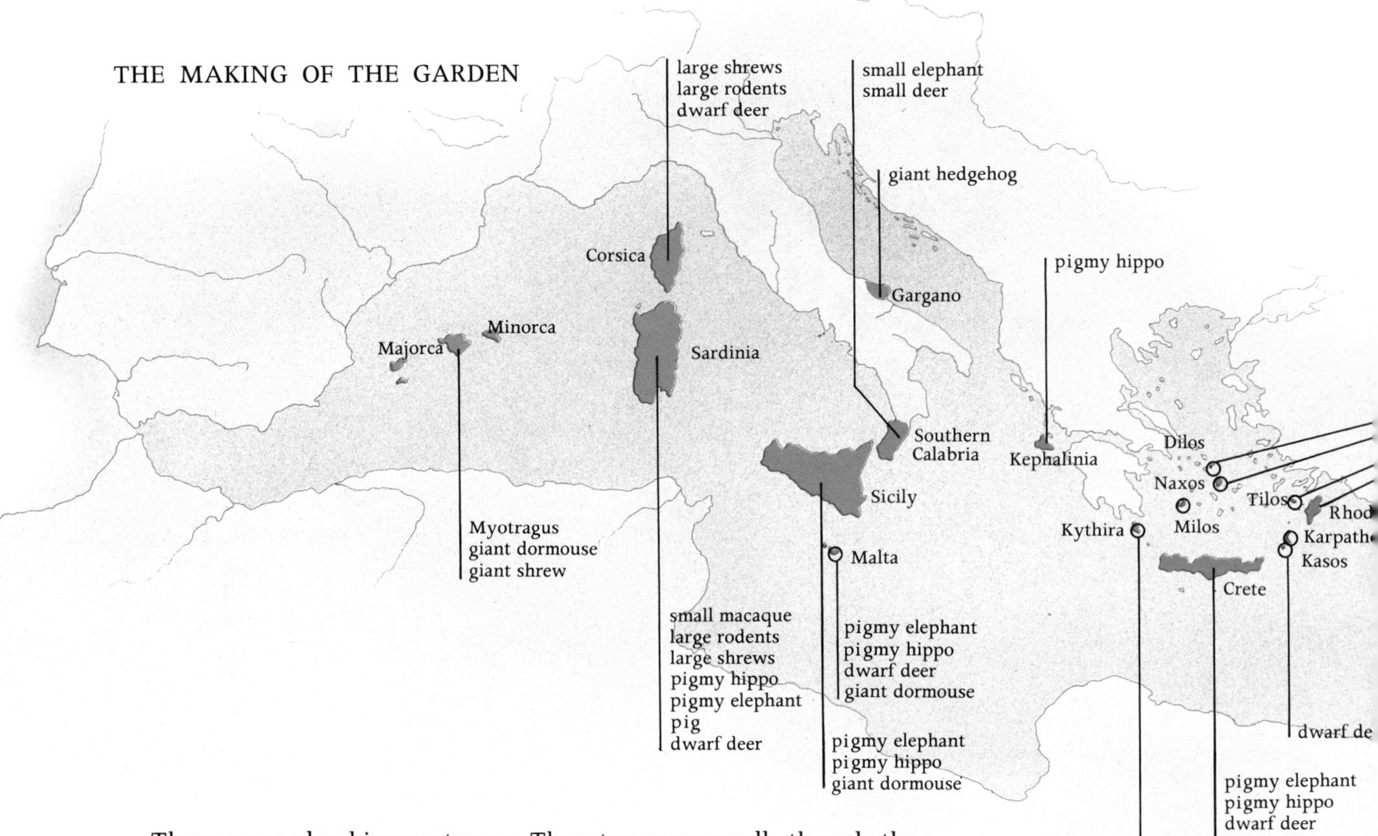

There were also hippopotamus. They too were small, though the difference between them and their full-sized continental ancestors was not as great in Malta as in the case of the elephants. But they had changed in another way. The joints of their limb-bones suggest that they were better adapted to moving about on land than to swimming. That must have suited them well, since they were living on an island where there were no large rivers. But again, the absence of enemies was probably a significant factor. African hippos feed on land and at night. During the day, they protect themselves from attack by lions by taking refuge in the rivers. With no lions present on Malta, the hippos were able to spend most of their days walking rather than swimming.

Deer also lived on Malta. They too were dwarfs and their legs had also changed in character, being proportionately shorter with some of the smaller bones fused together. This suggests that these deer were not such swift runners as their continental relatives, but had instead become more skilful at picking their way over the rocky terrain.

Other Maltese animals, however, had become giants. There was a huge swan with a nine-foot wing-span which, judging from the small size of the breast bone to which its flying muscles were attached, was probably almost flightless, a dormouse the size of a large rabbit, and tortoises fully comparable in size to those living today on the Galápagos and on Aldabra in the Indian Ocean. Birds escape their land-bound enemies by flying, rodents by taking refuge down holes and crevices. Both these factors normally limit their size. But in the absence of all enemies, both animals are free to grow large. The swans

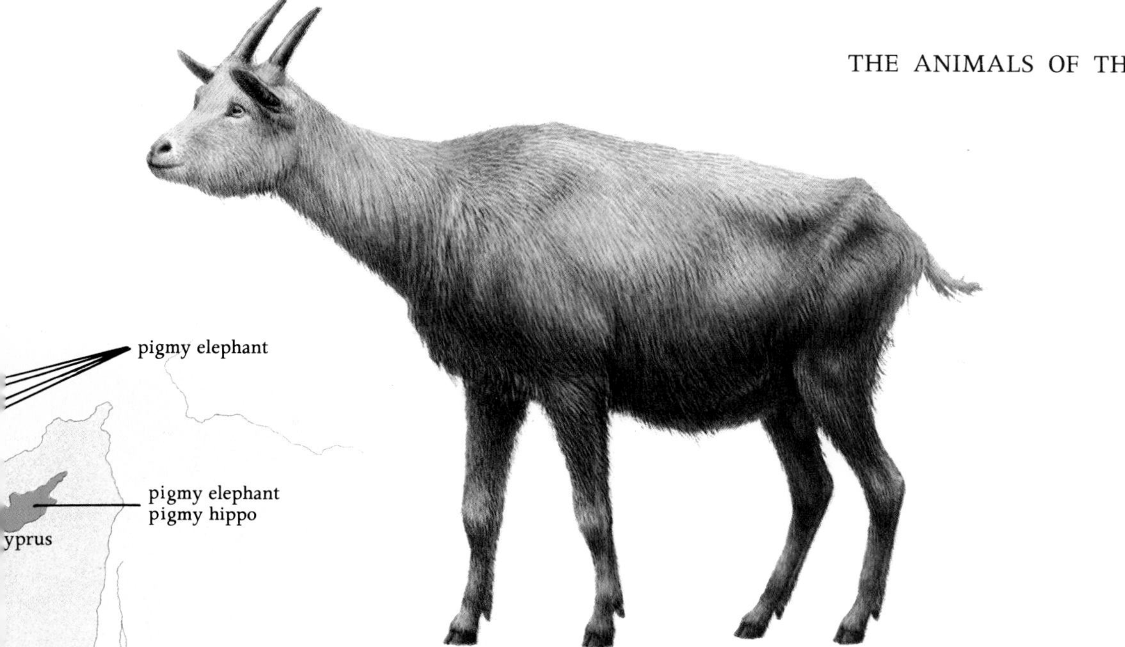

The strange animal of Majorca
When people arrived on Majorca six thousand years ago they found Myotragus, a kind of dwarf antelope. Its legs were short, reducing its speed but improving its agility. Its eyes were not on the side of its head, like those of an antelope that needs to watch for predators, but on the front, giving it stereoscopic vision valuable in climbing. We know that the people ate it (many of its bones show signs of butchering), and may have kept it in pens; but eventually it died out around four thousand years ago.

probably had relatively few problems with food, since little else competed with them for their favoured diet of water plants, but this would not have been the case as far as the rodents and tortoises were concerned. Competition for limited food supplies, it seems, has a different effect on small species than it does on large ones. Individual undersized elephants may well survive when bigger ones starve; but a big mouse or tortoise is likely to be more successful at claiming a share of scarce grazing than a small one.

The inhabitants of other Mediterranean islands were transformed in a similar way. Pigmy elephants and hippos occurred on Sardinia and Sicily, Crete and Cyprus. Corsica had giant rodents and dwarf deer. Sicily and Majorca also had giant dormice, similar to the species on Malta. A small island that once existed off the east coast of Italy (now attached to the mainland) was the home of a giant hedgehog, and in Majorca and Minorca a strange antelope evolved that had huge front gnawing teeth like those of a rat.

Most of these species appeared long after their islands were isolated. Although there may well have been colonisations between closely grouped islands such as Majorca and Minorca, or Sardinia and Corsica, it is not likely that they occurred between the more widely separated island groups. The odd animals on each of them almost certainly evolved independently. The pigmy elephants of Cyprus resemble those of Sardinia, fifteen hundred miles away, not because they are directly related, but because both are descended from similar full-sized ancestors and, faced with similar environmental pressures, both evolved in a similar fashion.

These strange creatures survived until comparatively recent times. Some of their bones are no more than about 8,000 years old. But while they were evolving, many other changes were taking place on the mainlands to the north and south of the islands.

THE CHANGING CLIMATE

At the time that the Mediterranean dried out, six and a half million years ago, the lands to the north and the south of it were relatively warm, well-watered and consequently covered by forests. Pine, olive and juniper grew on the lowland plains. At higher altitudes, those species were augmented by cedar and fir. Some idea of what those upland forests were like can be found in their remnants that today still linger, six thousand feet up in the Atlas Mountains of North Africa. Immense cedar trees, a hundred feet high and a thousand years old, dominate the forest. Among them grow maritime pines, evergreen oaks, holly and ivy, hawthorn and walnut. Bracken and peonies grow on the ground, and through the trees fly many birds familiar to Europeans, such as woodpeckers and tits, robins and ravens. But there are also macaque monkeys and, at the time of the drying, there were elephant, rhinoceros, and lions.

In the millennia that followed the refilling of the Mediterranean, the world warmed somewhat. The Atlas Mountains, which stretch across most of the width of Africa, acted as a barrier against the advance of the Sahara, and the forests of the North African coast not only flourished but extended around the Mediterranean, eastwards to the Caucasus and northwards as far as southern England.

But a change was coming over the whole world. The planet was cooling. The reasons that it did so are still not understood. The process had started around three million years earlier. Cold and relatively warm phases alternated; but it was not until 300,000 years ago that northern Europe became deeply affected. Sheets of ice spread down from the north, farther and farther, until they reached southern England and central Germany. Glaciers grew from the snow-fields capping the Alps and the Pyrenees, and ground their way far down the valleys and on to the surrounding plains. As the ice advanced, so the forests, and the animals that lived in them, were forced to retreat southwards. The Mediterranean region was sufficiently far south to remain unfrozen. Indeed, the lowland ice sheets did not get closer than about five hundred miles. Nonetheless, they had an important effect on it. As more and more of the world's water accumulated on land in the form of ice, so the level of the world's seas began to drop. It never fell so low that the rock barrier between Spain and Morocco was exposed again, but land bridges appeared between Corsica and Sardinia, Majorca and Minorca, and several of the Greek islands were linked into one. The arms of the Mediterranean that separated these islands from the mainland also narrowed significantly, and this may well have been the time when the first elephant and hippo swam or drifted across the Sea to colonise the islands.

Eventually the cold became so intense that most of southern Europe turned into a chilly tundra of grass and sedge, with only a few stunted

Mountains of the southern shore
The Atlas Mountains stretch for thirteen hundred miles across Tunisia, Algeria and Morocco, and form a barrier between the Mediterranean world and the Sahara. Several rise over thirteen thousand feet. Their northern flanks get good rainfall and were once covered with magnificent cedar forest. Today only a few tracts of it remain.

birch, dwarf willow and juniper reaching any height. Only on the southernmost shores of Spain, in the Middle East and the North African coast did the cedar forests survive.

The birds of those forests, however, were regular visitors to the European tundra. When summer came, the ground there unfroze, the small trees put out their leaves, and swarms of insects hatched out from the waters of the bogs. Rodents that had spent the winter in hibernation, woke up and began to scuttle through the grasses and sedges. Such things are good food for birds, and there were far more of them than could be consumed by the few birds that were hardy enough to live there all the year round. So birds from the African shore – insect-feeders such as warblers and swallows, rodent-hunters such as hawks and eagles, collectors of swamp-life such as storks – developed the habit of travelling northwards each spring across the Sea to nest and rear their young in the midst of this rich, if temporary bonanza. When the winter returned, the ground refroze, the food supply disappeared, and the birds flew back again to Africa, taking their young with them.

Several times Europe warmed, the glaciers retreated and the forests spread back into northern Europe, only for the cold to return and the ice to come south again; but ten thousand years ago the world became generally warmer, and the end of the Ice Age approached. As the ice sheets withdrew, so the feeding grounds around their southern margins moved northwards. But the birds did not abandon their habit of nesting and feeding there every summer, even though each year they had to fly farther and farther to reach them.

They still make that journey today. The majority of them travel by way of the Middle East, around the eastern end of the Sea. Storks, eagles and hawks that breed in central and eastern Europe fly up over Israel in huge numbers. One ornithologist has estimated that about a million of them make the journey each year. Those individuals of the same species that breed in western Europe do cross the Sea, but choose a short passage, either across the Gibraltar Strait or by way of the promontory of Cap Bon in Tunisia and across Sardinia and Corsica into northern Italy. Others go by way of Sicily, or travel to Cyprus, then into the smaller Greek islands and eventually into eastern Europe. But, surprisingly, a considerable number of these intrepid travellers, including such apparently frail creatures as small warblers, cross the Sea at its widest. They land, exhausted and starving, on the farther shore, and have to spend several days resting and feeding before they can continue with their journeys.

The increasing warmth has, within very recent times indeed, induced new species to join those that have been making these migrations for millennia. Flamingoes are essentially birds of the tropics that

The journeys of the storks
White storks spend the winter in Africa. In spring, they fly to Europe to breed. At the height of their migration, thousands fill the morning sky (*above*) as they soar upwards in the thermals, in order to gain maximum height and so complete the next lap of their long gliding journey with a minimum of wing-flaps. They avoid long sea-crossings. Some go by way of Gibraltar, others fly around the eastern end of the Mediterranean (*right*). Individual birds habitually follow the same route year after year: young birds hatching in one German village will return across Spain, whereas those from a nest only a mile or so away are quite likely to travel back by way of Istanbul.

find their food in shallow salty lagoons and there build the mounds of mud that serve them as nests. The greater flamingo is particularly adventurous, and lives not only in Africa but in Central America and western Asia. It has occasionally bred on the southern coast of the Mediterranean in Tunisia, and makes regular journeys across the Sea to nest in the lagoons of the Camargue, around the mouth of the Rhône. Here, it seems, it is at the very edge of its range. If a summer is cold and wet, it may not nest there at all. But if the climate of Europe were to warm a little there could be permanent colonies of these spectacular birds in many places along the northern shores of the Sea.

The fastidious flamingo
The greater flamingo (*right*) feeds only on small invertebrates that flourish in the mud at the bottom of shallow salty lagoons. It finds such places all round the Sea, on both northern and southern shores, but only in a few of them are the conditions exactly what it requires for breeding.

The tolerant stork
The white stork eats almost any small animal it finds, though it is particularly fond of fish (*above*) and frogs. It seems indifferent to human disturbance and regularly builds its nests on house chimneys and telegraph poles, on which it carries on its courtship and mating (*right*).

SPRING FLOWERS

Seeds are excellent water-tight, heat-resistant, fungus-proof capsules in which the genetic essence of a plant species is able to pass the rigours of the baking rainless months of the Mediterranean summer. The annuals rely on this device altogether, for they die as soon as they have seeded. The species of vetch (*1*), chrysanthemum (*2*), toadflax (*3*), lupin (*4*), catchfly (*5*) and pimpernel (*6*) shown here all use this technique. The genetic change that causes them to do so is, seemingly, easily acquired, for each of them has close relatives that are perennial and live for several years.

1

3

4

5

2

6

SURVIVING THE SUMMER

Today, around the European shores of the Mediterranean, the hostile season of the year is not the winter, as it once was, but the summer. The danger for animals and plants living there comes not from frost and bitter winds, but from heat and the potentially lethal drought that accompanies it.

Some small plants manage to avoid the summer altogether. They are annuals, such as some of the pinks and campions, mignonettes, stocks and vetches. They concentrate their entire lives into half the year. When the autumn rains come, they sprout leaves and grow quickly. By spring, they are in flower. With the advent of summer, they die and leave their next generation to survive the rigours of the hot months to come as scattered seeds wrapped in water-tight skins.

Iris and orchids, cyclamen, narcissus and asphodel come into bloom at around the same time and also, seemingly, quickly disappear. But individually these plants are longer-lived, for they have bulbs, or swollen roots, below the ground in which they store the moisture that they have collected and the food they have made. This reserve enables them to sprout their leaves and to start making more food as soon as there is sufficient moisture available in autumn so that they have the maximum available when the time comes to produce flowers and set seed at the end of the season. Some use their underground reserves to produce flowers even before their leaves, as soon as the rains arrive. Which species adopts which practice seems almost arbitrary. Some species of iris, crocus and cyclamen, for example, flower in autumn, whereas others do so in spring. But by May, the annuals have died, the bulbous plants have withered, and flowers have largely disappeared from the Mediterranean.

The shrubs, however, are for the most part evergreen. Growing leaves consumes a lot of energy, and wasting this by allowing them to shrivel and die represents a real loss to a plant. On the other hand, retaining leaves throughout the hot summer risks the loss of precious water by evaporation through the leaf pores. Evergreen shrubs, therefore, have to have methods of preventing this disaster. Most of their leaves are thick and leathery with waxy surfaces and very few pores. Many have a grey or silvery colour which reflects the heat of the sunlight. Thick mats of hair cover the surfaces both above and below, impeding the circulation of air over the leaves, and therefore the amount of water vapour lost, and many curl up so that the area exposed to the wind and the sun is reduced to a minimum.

Several shrubs grow special leaves for the summer that are even smaller, hairier and waxier than their winter leaves. This is a somewhat mysterious tactic, for these summer leaves reduce water loss from their pores still further by stopping gaseous exchange and food production altogether. Their value to the plant is therefore by no means clear.

SPRING FLOWERS

Plants have several different methods of storing, from one season to the next, the starches and sugars they manufacture. Lilies such as sea-squill (*1*) do so in a bulb, a modified shoot surrounded by short swollen leaves. Cyclamen (*2*) and gladiolus (*4*) have enlarged underground stems called corms; and asphodel (*3*) and iris (*5*) use tuberous roots. The yellow bee orchid (*7*) is one of over fifty species of that family that live around the Mediterranean, all of which grow on the ground and develop tubers from their roots. Those belonging to the genus *Orchis*, after which the whole family is named, such as the Anatolian (*6*), Provence (*8*) and butterfly orchids (*9*), develop a pair of small ovoid tubers which is why the Greeks gave them the name 'orchis', that being their word for testicles.

1

2

3

4

5

6

7

8

9

The leaves of the Mediterranean shrubs, however, have other functions apart from food-making. Thyme and sage, origanum and rosemary manufacture oils in their leaves. These, as they slowly vaporise in the summer heat, are responsible for the rich and characteristic fragrance of the Mediterranean summer. Almost certainly they assist in reducing water loss, for they evaporate much less readily than water, and coat the leaves with a thin protective film. But they may also provide the plants with a defence. Most grazing animals find them distasteful and do not therefore damage the shrubs. But such leaves may also be weapons in the battle to claim precious moisture from the parched land. The oil vapour from sage leaves is known to retard the growth of seedlings. As a consequence, individual sage plants are usually surrounded by an area of bare earth. Oil from thyme has this effect even on thyme seedlings and so helps to ensure that individual thyme plants are widely separated from one another. So it may be that, even though the summer leaves of so many of these shrubs do not manufacture food, they nonetheless serve their owners very effectively in defending their position during the most difficult time of the year.

Where the land is rocky and the soil is thin or almost non-existent, smaller lower shrubs grow, producing the kind of country known in France as the *garrigue*. Small bushes – lavender and sage, rosemary, thyme and heather – sprout from the bare rock, their long sinewy roots reaching deep into the crevices to extract nourishment from the few particles of soil that have accumulated there. Where the soil is just a little thicker, the vegetation can grow higher and more densely. This is the *maquis*, with oleander and broom, gorse, myrtle, laurel, arbutus and rock-rose.

The fragrant maquis in Spain
Beneath evergreen cork oaks grow gum cistus from which laudanum was once made, arbutus with strawberry-like fruits, and lavender (*right*), the leaves of which contain perfumed oil.

Summer leaves
In summer, sage (*far left*) replaces its long winter leaves with small ones from which little moisture evaporates, and which contain an oil that animals find distasteful. The thorny burnet (*left*) also grows tiny summer leaves. They lack oil, but instead the plant protects itself from grazers with a mass of spines.

Tall trees also grow along these shores. Only a few centuries ago, they were much more widespread than they are today. They too, for the most part, are evergreens and include several species of pine, holm oak and cork oak, juniper, cypress and wild olives. All have leathery leaves protected by thick waxy surfaces.

The hot, dry summer poses similar problems to animals as it does to plants. A few insects have developed ways of feeding during this harsh time. The cicada, whose long shrilling calls fill the summer days, has a long proboscis like a stiletto with which it is able to pierce the bark of trees and suck out the sap. Other insects, that are not so well equipped, come to take advantage of the cicada's skill. As it drinks, so ants, wasps and flies gather around it, collecting such tiny trickles as may escape from the tree's wound.

But cicadas are exceptional. Most animals have to take steps to shelter during this oppressive time. Whereas in northern Europe, the winter cold forces many of them to go into a form of inactivity known as hibernation, here in the south the summer compels aestivation, a similar retirement from the scene. Snails secrete mucus which solidifies and seals off the entrance to their shells so that no moisture can escape, no matter how hot the summer. They will remain so until the rains of autumn arrive to dissolve the seals. Earthworms burrow deep into the soil and curl up in small chambers where there is still some vestige of moisture. Ladybirds have hard impermeable skins and are in little danger of drying out. They feed, however, on greenfly which are easily dehydrated and disappear during the summer. With nothing to eat, the ladybirds congregate in thousands and pass the summer huddled together beneath rocks, though why they should choose to do so communally is a mystery.

The Jersey tiger moth also behaves in this way. It is a brightly coloured insect, with black forewings blotched with white or pink, and hind wings that in some individuals have paled to yellow and in others intensified into a bright red. As a species it is found over much of Europe, ranging as far north as southern England and the Channel Islands, from whence it gets its English name. Over most of its range it remains active throughout the summer, like any other moth, but in the hotter drier parts of the Mediterranean the summers are too harsh for it to do that. So in a few places, and most spectacularly on Rhodos, the most easterly of the Greek islands, it has adopted a special form of aestivation.

Throughout the warm moist winters, the caterpillars are common all over the islands, feeding on a wide variety of vegetation. By May, they are fully grown and as spring changes to summer they turn into adult moths. Over the next hot, rainless weeks they all make their way to one small wooded valley where a stream runs throughout the

Summer strategies
The first four years of a cicada's life are spent as a grub below ground feeding on plant roots. It emerges as an adult (*right*) at the height of summer. Its life is then full of activity as it energetically calls throughout the heat of the day to attract a mate, simultaneously drinking tree sap through its long proboscis thrust through the bark. It lays its eggs and then, after six weeks at the most, it dies.
For snails (*far right*), on the other hand, summer is a time when everything must stop. Their soft moist bodies would lose far too much moisture in such heat. They therefore retreat into their shells, seal the entrance with a water-tight plug of mucus, and gather together to sit out the hot months in complete immobility.

A million moths

Petaloudes Valley – the name means butterfly – is the one place on the island of Rhodos which remains relatively cool and moist throughout the hot dry summer. A stream, fed by an unfailing spring, runs down the valley and supports a dense grove of trees. There, in the humid shade, a million insects assemble to wait for the return of the cool months. In spite of the valley's name, they are not butterflies, but moths, the Jersey tiger. Several other assemblies are known elsewhere in the Aegean islands and southern Iran, but none can rival the spectacular numbers of the Rhodos gathering.

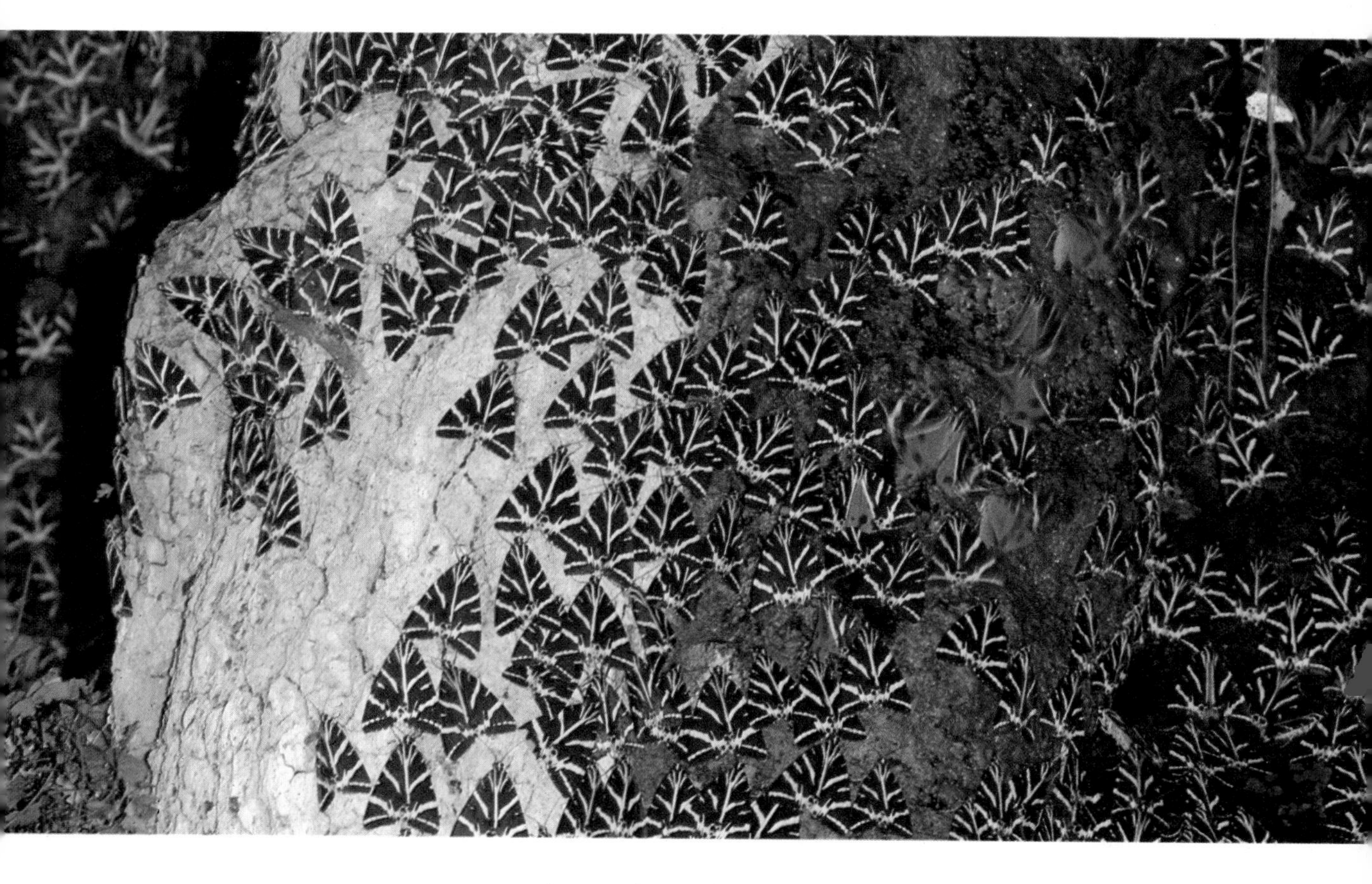

year. Thousands upon thousands arrive until they cover every stone and tree. The most favoured position of all is a moss-covered rock cliff beside the valley's single waterfall. There the air is most humid, and there the moths cling in a continuous carpet.

They fed well enough when they were caterpillars to build up stores of fat. This they must now eke out if they are to stay alive throughout the summer months. They reduce their bodily processes to an absolute minimum. Most of the time, they remain completely motionless. The humid air of the valley prevents their bodies from drying out, and the perches they prefer are seldom reached by direct sunlight, so they do not over-heat. At night some may flutter up into the branches above, but as soon as the sun returns they fly feebly down again into the moist shade. Nothing attacks them: their bodies contain distasteful poisons and their vivid colours warn birds of the fact. If they are disturbed by the movement of an intruder or the call of a bird, they may struggle into the air, but for the most part they remain absolutely still, saving their energy.

As summer draws to an end, they once again begin to stir. And now they mate. Some have been seen to drop into the stream during their courtship, complete their copulation under water and clamber out afterwards, apparently unharmed. Powered by the last remains of their fat reserves, they fly out of the valley and disperse over the island which is now, once again, cool and moist. Now, at last, they feed once more, sipping nectar from heather and ivy flowers. The females in one last bout of activity lay the eggs that, with the coming of the winter rains, will hatch out into a new generation of caterpillars. And then they all die.

Reptiles, however, relish the summers. Their skins are watertight, so the danger of dehydration for them is a negligible one. Furthermore, they gather the heat they need to activate the chemistry of their bodies directly from their environment instead of using food to generate it as mammals do, so warm days give them more energy rather than robbing them of it. In consequence, reptiles of one kind or another are common. Tortoises are found around most of the Sea's coastline. Lizards, some small and brown, others brilliant green with blue spots on their flanks, sun themselves on rocks, snapping up flies and ants. Some two dozen different species of snake live around the northern shores, the biggest of which, the Montpellier snake, can grow to seven feet long. Although like several others it carries poison, the fangs with which it delivers it are placed right at the back of its jaws, so it has to take its prey well into its mouth in order to inject any venom. It can do that with lizards, other smaller snakes and nestling birds; but it cannot do it with larger creatures, including human beings.

Few of the land mammals of Europe, however, find such dry heat tolerable, and most restrict themselves to the mountains or the cooler lands to the north. Hedgehogs, rabbits and hares, foxes and mice are the most abundant of them, and even they are relatively scarce. The mammals of Africa might be expected to find such conditions more congenial than their European relatives do, and certainly southern Europe does provide a home for a few species of African origin.

One, the macaque monkey that lives on the Rock of Gibraltar relishes the hot conditions; but it originally came to Europe long before the present climate was established, about five million years ago, soon after the dried-out Mediterranean refilled. At that time, the ancestral macaques colonised the eastern end of the Sea and spread into Asia, where they diversified into the many species that live there today. Others went west into Europe, as we know from their fossil remains, and even spread as far north as Britain.

Macaques are among the hardiest of monkeys, the only ones that are able to endure real cold. Those that still live today in their ancestral home, the Atlas Mountains of North Africa, have to endure

REPTILES OF THE MEDITERRANEAN

Warm climates suit reptiles, which draw the heat for their bodies directly from the sun, and the Mediterranean shores of Europe are richer in them than any other part of the continent. There are three species of tortoise of which the spur-thighed (*3*) is the most widespread, being found in Greece, Turkey and Italy as well as along the African shore. There are two dozen or so species of snakes. One of the biggest is the Montpellier snake (*2*) which can grow to seven feet long. The sand boa (*4*), a member of the family which contains the biggest of all snakes, is itself less than two feet in length. The most handsomely marked is perhaps the leopard snake (*1*) that lives in Greece, southern Italy, Sicily and Malta.

1

2

3

4

REPTILES OF THE MEDITERRANEAN

Lizards from five different families live in Mediterranean Europe. The Lacertids are the most abundant and widespread. The most spectacular of their thirty-six species is the ocellated (*3*) which grows to nearly two feet long (though two thirds of that is tail) and eats a wide variety of food – insects, fruit, birds' eggs and even small mammals. The agama (*5*) is the only member of that large African family to reach Europe and has established itself in eastern Greece. Skinks are ground-livers with a tendency to burrow and wriggle rather than use their diminutive legs. Five, including the ocellated (*4*), live in Europe. Geckos have adhesive pads to their toes which enable them to cling to walls, and are nocturnal. The Moorish gecko (*2*) like the other three species that occur in Europe, habitually stations itself beside lights in order to snap up the insects that are attracted there. The chameleon (*1*) is found in southern Spain and Greece, but probably reached these outposts to its primarily African range with the help of man.

1

2

3

4

5

winters that regularly cover their cedar forests in snow many feet deep. They are able to survive because they develop each winter a particularly thick coat. They have also, during the course of evolution, lost the long tail that other African monkeys possess, which is why they are sometimes erroneously called Barbary 'apes'. Clearly such a long appendage would be a handicap in very cold weather: a lot of body heat would be lost from it, and it would be very vulnerable to frost-bite. This tendency to lose their tails is also found among Asiatic macaques. The Japanese snow monkey, which is one of them, is tail-less. So too is the Celebes black ape, which evolved in the mountains of that island.

The European macaques, hardy though they may have been, could not withstand the rigours of the Ice Age when the glaciers began to spread down from the north 300,000 years ago. They were driven south until they survived only in Andalucia, the southernmost part of Spain. The origin of today's Gibraltar troop is uncertain. They could be the last descendants of that ancient European population. Records of them, however, only go back to the eighteenth century and it is possible that those were descended from animals that were brought across the Straits as pets by human beings and subsequently went wild. Around that time, a saying began which maintained that if the monkeys left the Rock, so would the British. Since then, the British authorities in Gibraltar have imported several batches of new stock from Africa to make quite sure that such a thing did not happen. So even if the colony was once a natural one, it is no longer so.

Other African animals that today live on the northern shores of the Sea have certainly been assisted in their spread by man to suit his own purposes. Genets, small, close-furred, lissom cats that under their own volition have advanced around the eastern end of the Sea into Israel, were introduced into Spain to help in the control of rats, and have since gone wild. The same thing happened to the Egyptian mongoose, which is adept at catching not only rats but snakes as well, and which lives wild in the Camargue.

One African mammal can claim, beyond contradiction, to have invaded Europe entirely by itself. It is a fruit bat.

Typical fruit bats are very different from the small insect-eating, cave-dwelling bats common in northern Europe. They are essentially a tropical group; they are much bigger, their bodies varying in size from that of a rat to a small fox; they roost out in the open in trees; they navigate, not by echo-location as the insectivorous bats do, but by vision, and they have large eyes that enable them to do so. They live, not on insects, but as their name suggests predominantly on fruit.

The Egyptian fruit bat, however, is an exception. It is certainly quite large, with excellent eyes, and is a fruit-eater. But it has quite

The cool forests of North Africa High on the northern slopes of the Atlas Mountains, where in winter snow lies many feet thick, forests still survive that are very like those that covered the European shores of the Sea at the end of the Ice Age. Robins, tits and nuthatches fly among cedars, hawthorn, evergreen oak and yew. On the ground grow cranesbill, chickweed, asphodel and other woodland plants still common in northern Europe, as well as a few specialities such as the Atlas peony (*right*) which grows nowhere else.

The Mediterranean's monkey
The Barbary macaque (*left*) still lives in troops in the high cedar forests of Morocco. This is the source from which new recruits have been taken to maintain the species's colony on the Rock of Gibraltar.

An African carnivore in Europe
The common genet (*above*) is found in forests all over the Iberian peninsula and southern France, as well as throughout Africa. But it is rarely seen, for it is only active at night, and runs through the branches of trees with such agility that it is almost soundless.

independently developed its own kind of echo-location, producing the sounds it needs for navigation by clicking its tongue and drawing its lips back in a grin, so that the sound is emitted from the sides of its mouth. This skill has allowed it, uniquely among fruit bats, to navigate in the dark. The cold Mediterranean winters would kill fruit bats if they had to roost hanging in trees. But the Egyptian fruit bat, by virtue of its navigational skills, is able to take refuge from the cold in deep dark caves; so it alone has been able to establish colonies outside the tropics. Great numbers of them now live in Cyprus. During the summer, they move out of their caves, roost in trees and feed largely on figs. What they find to sustain themselves throughout the winter has yet to be discovered.

One other African mammal made the journey to Europe. It arrived several million years after the macaque monkey, but long before the Egyptian fruit bat, or any other African mammal living in Europe today. Eventually, it colonised not only all the lands around the Mediterranean, but it spread northwards over the whole of Europe, eastwards into Asia and even far beyond that. This was that most ubiquitous of all mammals – man.

An African mammal in Cyprus
The huge eyes of the Egyptian fruit bat (*left*) enable it to navigate excellently on moonlit nights, but it has also developed a sonar technique. This is not nearly as sophisticated as that of insectivorous bats but it is unique among the fruit bats, and enables it to fly into deep caves (*right*) where there is no light whatever.

THE ARRIVAL OF MANKIND

The hunting grounds of prehistory
The limestone valleys of eastern Spain (*right*), now so rocky and bare, were 10,000 years ago covered by thick forests rich in game. Beneath the overhanging cliffs, bands of hunters regularly sheltered. The animals have now gone, but the drawings the hunters made of them on the rock walls still remain.

Fossil evidence has now made it quite clear that the genus *Homo* originated from ape-like ancestors in Africa some time around two million years ago, long after the Mediterranean had refilled. These early men began to spread northwards until eventually they reached the shores of the Sea. Remains of them found in Algeria show them to have been more stooped, heavier-browed and more massively jawed than any human being alive today. They have been called Neanderthalers. By what route they crossed the Sea is debatable. Some certainly travelled around the eastern end of it, through the countries of the Levant and into Europe by way of Turkey. But others may have moved from island to island across the middle of the Sea. By a million years ago, such people were living in southern France.

In ecological terms, the Neanderthalers were predators, like bears or any other hunting animals in the forest. Like them, they were not very numerous compared with the animals they preyed upon, cattle, reindeer and horse. Like their fellow hunters, they made little attempt to modify the world in which they lived. They simply accepted the bounty of the forest as they found it. In Spain, the remains have been found of one of their hunts in which a herd of elephant were driven into a bog and thirty or forty slaughtered. Crudely worked flint choppers lie among the dismembered bones. In France, near Nice, archaeologists have unearthed one of their hunting shelters, the oldest human structure known. It is an oval hut, made of branches with a low wall of pebbles on its windward side. The thinness of the layers of refuse on its floor suggests that it was only occupied for short periods at a time, and pollen grains in the human droppings make it clear that these visits took place during the summer.

By about 30,000 years ago, the character of the human populations had changed. The skulls no longer had prominent ridges above the eyes; the jaws were less heavy. These people physically were very like us. The tools associated with their bones are more refined. Flint flakes have been more expertly worked into knife blades and spear points. Whether this new form of humanity had evolved in Europe, or whether it first appeared in Africa, has not yet been settled. Nor

The first human constructions
Careful excavations near Nice revealed the remains of a hunting camp built near the beach beside a forest of pine and oak around 400,000 years ago. One shelter was 45 feet long. It was probably covered with sprays of leaves to help keep out the rain.

is it certain whether or not they were descended from the Neanderthalers. The increasing abundance of their fossilised bones, however, shows that over the next few centuries, they became more and more abundant until, eventually, the Neanderthalers disappeared entirely and the newcomers alone survived. They were our direct ancestors.

A cave at Parpallo, on the east coast of Spain, just south of Valencia, provided these new people with one of their earliest Mediterranean homes. It lies fifteen hundred feet up in the limestone hills facing the sea. When archaeologists first began investigating it, the cave was full of soil which they had painstakingly to remove. On its stone floor, they discovered remains that they were able to date as being around 28,000 years old. Among the domestic refuse, they found simply worked flakes of flint that had served the people as scrapers, knives and arrow-heads. Ten feet above this level – and approximately ten thousand years later – the tools they excavated had changed in character. They were much more refined and elegant, and had been made with the sophisticated technique known as pressure flaking. The lightness and sharpness of the arrow-heads suggest strongly that the people were now using that powerful and accurate weapon, the bow, that was to serve man as his main hunting armament until only some five hundred years ago.

From the bones found in the same deposits as the tools, we know accurately what animals they hunted. Rabbit and ibex were by far the most common, and those animals were probably living on the mountain slopes near the cave. But there are also the bones of cattle, deer and wild horse, and for those the hunters must have gone down to the plains along the coast several miles away and have carried back the heavy carcasses, perhaps partially butchered, to their family in the cave. They also painted designs on small plaques of stone. Over three thousand of them have been found, some decorated with seemingly abstract patterns of zig-zags and swirling ribbon-like shapes. Others were executed at the same time showing horses and wild cattle.

Over the following centuries, the Ice Age slowly loosened its grip on Europe, and the character of the forests growing on the hills of southern Spain began to change, and with it, the animals that lived in them. By ten thousand years ago the climate at the time was very similar to that of today. The hunters who wandered through the forested valleys took to making temporary camps beneath the overhanging cliffs and spent hours recording on the walls the details of their lives in astonishing and vivid detail. Their paintings, executed in black, red and white ochres, are now varnished by a thin layer of calcite, deposited by water trickling down the cliff face. But a spray with water reveals them, like images developing on photographic paper. Groups of stick figures prance across the cliffs brandishing bows and arrows. Some flee from rampaging bulls. Figures lie wounded on the ground, with arrows in their sides. Men climb trees, bag in hand, to gather honey from a nest of wild bees, with the angry insects buzzing around their heads. Some wear masks and dance. Others, more forebodingly, have assembled in groups and appear to be fighting.

These vivid animated images show clearly that the people living around these western Mediterranean shores ten thousand years ago were still living by hunting and by gathering roots and fruits from the forest, just as they had done ever since they first arrived on the shores of the Sea. But at the other end of the Mediterranean, other human communities were developing a different way of life. They were building permanent homes and settling together in villages. And they were devising ways of exploiting the animals and plants around them that would eventually transform the whole world.

Two hinds from a hunter's shelter
Valltorta in eastern Spain contains many prehistoric rock paintings. Shielded from the infrequent rains by the overhanging cliffs, they have survived with astonishing freshness.

A hunting scene from Spanish prehistory
Hunters carrying bows, and two figures seemingly with feather head-dresses, are surrounded by goats, deer and bulls, three of which have had antlers added to their horns.

PART TWO

THE GODS ENSLAVED

Hunters, painted ten thousand years ago in the Valltorta, Eastern Spain.

THE SACRED ANIMALS OF PREHISTORY

The most formidable animal in the forests that grew around the Mediterranean ten thousand years ago was the great wild bull. It stood six feet high at the shoulders and weighed about a ton. It was long-legged and lightly built, though with a powerfully muscled neck. In colour it varied. Those in Europe were an intense black when adult, with a white streak along the spine and a white patch around the muzzle. Those in Africa were more reddish, and had a pale-coloured 'saddle' across the middle of the back. Their horns were huge – as much as three feet long, whitish in colour with black forward-pointing tips. And they were fast and aggressive. We can describe them so precisely because they survived in the forests of central Europe until the middle of the seventeenth century, and we have eye-witness accounts of them.

But even without this recent evidence, we know their dimensions from their fossilised bones, and their appearance from superb portraits of them painted by prehistoric man. While northern Europe was still gripped by the Ice Age, 35,000 years ago, people began to draw images of these bulls on the cave walls of France and Spain. The most spectacular so far discovered are in the cave of Lascaux in central France. There, in the main chamber, five huge bulls span the ceiling. The biggest of them is seventeen feet from head to tail. Some are drawn in outline only, others are fully finished and coloured. Around them are portrayed the cows. In life, these were about three-quarters of the size of the bulls, and a uniform brown or reddish colour. On the walls of Lascaux, they are comparatively tiny, and dwarfed by the massive bulls around which they cavort and leap. As well as cattle, there are other animals – deer, bison, ibex and, in numbers that rival the cattle, horses. Elsewhere in the hundred or so painted caves that have been discovered, there are pictures of mammoth, boar, rhinoceros and, rarely, fish. But the two dominant animals are horses and cattle. Between them, they constitute over half of all the paintings.

Much has been written about what was in the minds of the artists who decorated the caves. An early suggestion was that the act of painting played a part in rituals designed to bring success in hunting, and to ensure the continued fertility of the creatures on which the people depended for food. There are several pieces of evidence that support such a theory. For one thing, cows are occasionally shown with swollen bellies as though in calf, and mares as though in foal, which suggests that the painters were particularly concerned with fertility. Some images have V-shaped designs drawn across or close by them that may perhaps symbolise spears or arrows. One bison is shown so badly wounded that its entrails hang from its gashed stomach. Furthermore, all the animals represented are those that we know to have been eaten from the evidence of bones found on living

The Great Hall in Lascaux showing one of the great bulls and the heads of two others.

sites. Animals that were not eaten but which were nevertheless common in the forests hardly ever appear. One or two designs may perhaps represent dogs, but their identification as such is questionable. The same is true of birds. Human figures are extremely rare. There are no flowers, no trees, no landscapes.

In recent years, this motivation has been questioned on the grounds that, at the time the paintings were made, game was so abundant in the forests that magic was hardly required to find it. Such criticism ignores the fact that, no matter how common the animals were, bringing down such a large and powerful creature as a horse or a bull with nothing more than a flint-tipped spear or an arrow was both difficult and dangerous. Hunters could surely be expected to do everything in their power to ensure their success and safety in such encounters – and, after all, a harvest festival in an English country church does not imply that the crops are poor.

A later theory proposed that the animals fall into two separate groups. One, headed by the horse, represented the male principle (irrespective of the sex of the individual animals); and the other, which included cattle and bison, the female. The placing of different species within a cave, it claimed, had a magical significance, so that the whole cavern formed a kind of unified temple where people invoked the two aspects of sexuality, and performed rites of fertility.

We shall never know, for certain, the precise significance of the paintings, but one thing seems beyond doubt: the making of them was of the greatest importance to the people. The caves where they were placed were not, in most instances, living sites but chambers far from the light of day. The artists must have been brave indeed to have ventured along those black tunnels, squeezing through narrow clefts or, where the ceilings are low, crawling on all fours, finding their way by the feeble light of a wick burning in a dish of oil or animal grease, or perhaps of a flaming faggot that soon would burn out. Once in these far chambers, remote from the world of sunshine, they must have laboured for many hours, painting their magnificent and often very accurate designs. So whatever their precise purpose was, it was not trivial. These pictures are no idle doodlings made to pass the time. The feeling they evoke, particularly when they are viewed not in the flat illumination of electric light as a visitor normally sees them today, but fitfully and dimly by the flickering flame of a lighted wick, is of deep reverence for the wild creatures with which the painters shared life and on which their own lives depended.

Whether or not the nomadic hunters, living many millennia later in the Spanish rock shelters and elsewhere around the western end of the Mediterranean, shared these attitudes towards the animals they hunted cannot be demonstrated. The lively scenes they painted

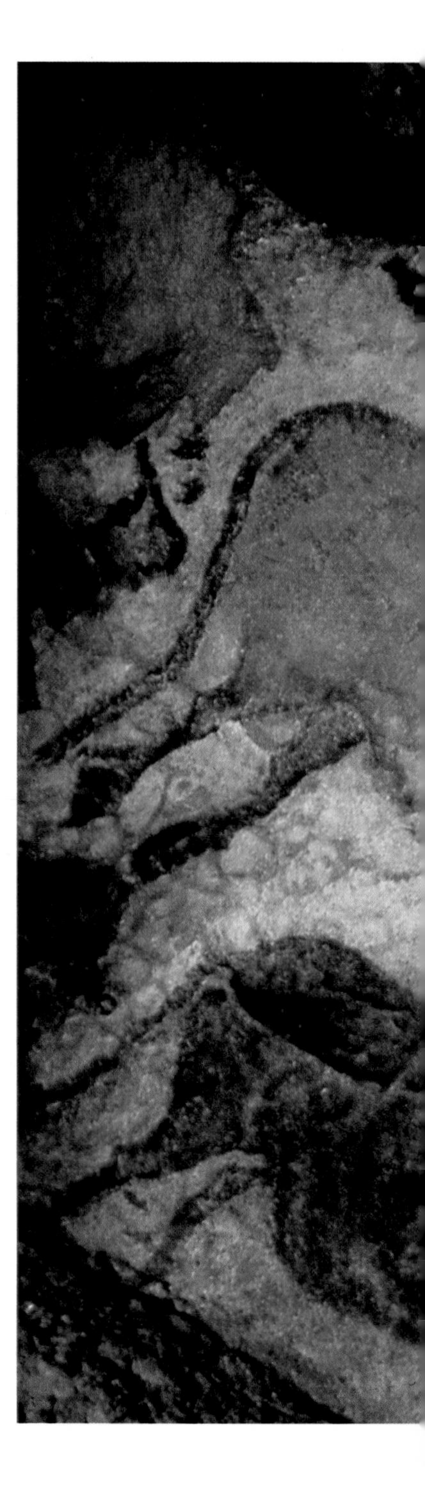

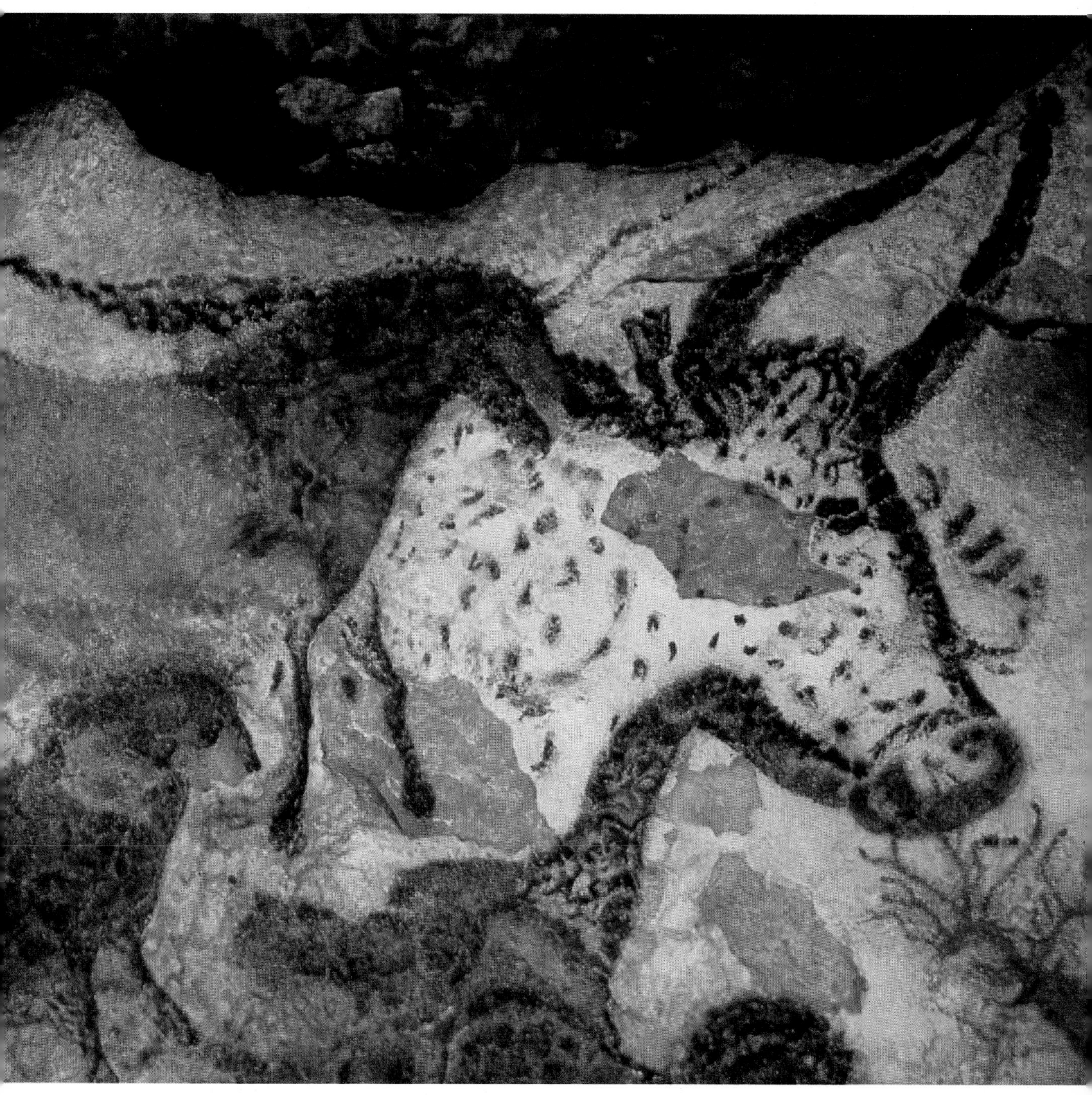

One of the great white bulls of Lascaux, with two horses drawn across it, and an antlered deer in front.

– mere sketches compared with the grandeur of the greatest of the French paintings – tell us a great deal about their everyday life, but shed little light on their religious beliefs. It is hardly likely that they were without gods, but we have no evidence of what those gods were. At the eastern end of the Sea, however, the bull, which had dominated the minds of men for so long in central France was still being revered.

There the people were beginning to abandon their nomadic hunting life. For many centuries past they had been gathering grass seeds for food. At first they had roasted them. Then they had ground them into flour, using pestles that they had originally devised for the crushing of ochres. Eventually they discovered that if they saved some of the seed and scattered it on cleared land around their settlements, the task of gathering fresh supplies the following autumn became much less arduous. They had begun to farm.

The seed of wild grass is difficult to gather. If it is collected before it is ripe, it is not very good to eat. On the other hand, if it is left ungathered for long after it has ripened, the seeds will have fallen from the stem so that they may be distributed by the wind or the feet of animals. So there may be only a few days when the harvest is possible. Seeds that remained attached for slightly longer periods would, inevitably, be the ones most likely to be collected. So one of the first effects caused by people gathering and resowing grass was to produce strains that retained their seeds longer than the wild forms. Eventually, this tendency went so far that the plants did not shed their seeds at all. Thus fully domesticated cereals – wheat and barley – lost their ability to distribute themselves and became totally dependent on man for their propagation and spread.

As the practice of sowing grain became more widespread, the land produced more food for human consumption, and the human population increased. Small family settlements grew into villages; and villages expanded into towns.

One of the earliest big towns to be discovered developed around 8,400 years ago at Çatal Huyuk in central Turkey, about fifty miles from the northern coast of the Sea. It covered some thirty acres and contained a population of some 6,000 people. They lived in tightly packed, flat-roofed, mud-brick houses into which they descended by way of a ladder through the roof. A third of all the rooms found by the excavators were shrines. And in them were images of bulls. Outlines of the animal in profile had been carved on the wall plaster. Bull's heads, modelled in clay protruded from other walls, their long spreading horns running as ridges along the wall surface. Small stylised heads of bulls, made of dried mud and fixed to the front of an altar, had been replastered and repainted as many as a hundred times,

First of the bull shrines
The people of Çatal Huyuk, one of the earliest towns so far discovered, worshipped the bull within their houses. This is a drawing of one of their shrines when it was first unearthed. Six pairs of bull's horns had been embedded in a clay bench with the remains of a seventh, set slightly higher, near the wall.

presumably during the repetition of rituals. In some shrines, the frontal bones of bull skulls, still bearing the bony cores of the horns, had been built into the walls. In one particularly elaborate example, seven pairs of wide-spreading horns had been mounted, one in front of the other, on a long clay bench. Sometimes the bulls were associated with a human female figure, apparently a mother goddess. Several other animals are represented in the shrines, either in effigy or by their bones – vultures, foxes, weasels, leopards and rams – but none of them rivalled in number or size the images of the bull.

There is no certain indication that bulls were sacrificed in these chambers. There were no remains of altars on which they might have been slaughtered, no stains of blood nor signs of burnt meat. Indeed, the buildings are so tightly packed together that it seems hardly possible that a bull could have been brought into any of the shrines. There were, however, many small offerings – eggs, stone tools old and new – buried in small pits beneath the floor of these chambers. All this suggests that the bull, though it was killed for food and its horns brought to the shrine, was not regarded as a creature to be sacrificed as an offering to some greater spirit, but was in itself an object of worship.

It is difficult to tell whether or not the people of Çatal Huyuk had yet domesticated the animal which played such an important part in their religious lives. One of the first anatomical effects of the domestication of animals is a reduction in their size. Why this should happen is not certain. Early cattle-breeders may, understandably, have chosen smaller animals to keep as captives, since they would be easier to manage and when necessary handle. Maybe captive animals had a less abundant and varied diet: they would have been penned up at night to prevent them roaming and to protect them from predators, and so they would have been unable to graze at night as they would have done in the wild. The horns of domesticated animals also tend to become smaller. That, at least, may have been due to a deliberate and very understandable selection on the part of the herdsmen.

Such anatomical changes are not detectable in the cattle bones recovered from the early levels of Çatal Huyuk; but many generations must pass before such tendencies become marked, so it may be that, even at this time, the townspeople were keeping cattle captive. They certainly possessed other domesticated animals.

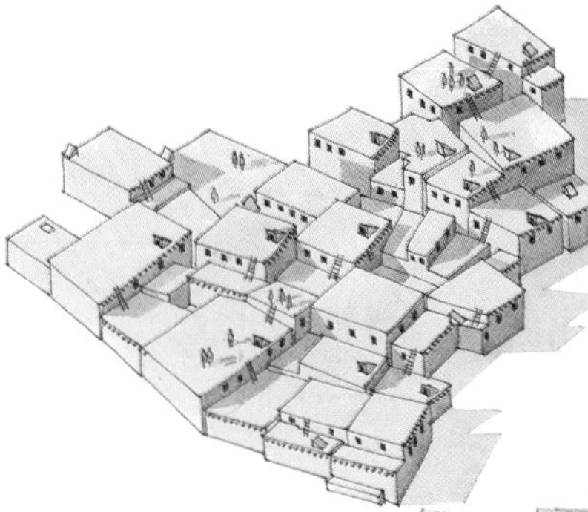

A section of the city
The houses, built of mud around a timber framework, were entered down ladders from openings in the roofs. Inside were store rooms and living rooms with stone querns for grinding flour. A third of all the rooms excavated, were shrines.

A bull sanctuary in Çatal Huyuk
The bull heads, modelled with clay, have been overpainted with designs, some many times. There are also goat heads, some set with actual horns, and pairs of women's breasts. Human skulls are placed on the benches, and the walls bear stylized designs of giant vultures attacking headless human figures.

The dog had been tamed centuries earlier. Even while the people were still nomads, they had recruited young orphaned wolves and possibly jackals as hunting assistants. The relationship was perhaps as much a partnership as a domestication. Men were helped in their tracking by the dog's super-sensitive nose, and in attacks on their quarry by the dog's sharp teeth. The dog, in return, took a share of the meat of the kill and gained the protection and warmth of a fire at night. That young dogs readily entered into such a partnership can be explained by the fact that they were themselves social animals, habitually hunting in teams, exchanging frequent signals with one another, and accepting the leadership of one dominant animal. In their new circumstances, that leader ceased to be a dog and was instead a man.

Sheep and goats, too, were living in tamed herds around Çatal Huyuk, and their bones, smaller than those of the wild form both in stature and horn size, have been found around several village sites in the Middle East dating from some ten thousand years ago. The animals from which these domesticated creatures were descended lived in Greece, Turkey and the lands to the east. No such wholly wild sheep survives there today. A species of sheep, the mouflon, does however live in the mountains of Corsica and Sardinia. Fossil remains have not been found there to suggest that it evolved on these islands, so it may well have been taken there by man soon after it was domesticated, and thereafter returned to the wild state. It provides, however, the closest approximation we have to the ancestral sheep. It is very different from the docile, long-suffering animal that lives on farms today. It is extremely shy and has remarkably acute eyesight. During the rutting season, the rams establish territories for themselves, announcing their claims by thumping their hefty horns against tree-trunks or rocks. The ewes with their lambs live in small separate groups.

Doubtless, the first stages of domesticating these animals occurred when people hunting the wild species caught and reared the kids and lambs whose parents they had killed. The young animals, growing up among human beings, did not have a fear of them; and eventually small herds remained conveniently close to the villages, and were claimed and protected by the inhabitants.

How people tamed the bull is much harder to imagine. Such huge beasts must have been extremely difficult to manage or even to keep penned inside an enclosure. Nonetheless, by six and a half thousand years ago, the wild bull had certainly been made captive, and from it are descended all the domesticated cattle of Europe today. It may have lost its freedom by this stage in history, but it had not yet, by any means, lost its sanctity.

THE ANIMAL GODS OF EGYPT

Cattle were also being kept around this time on the lush grasslands of the Nile delta. There, too, nomadic hunters were settling down to become farmers and herdsmen. Each of their settlements appears to have adopted an animal as a totemic god, and this was most commonly a bull. Several burials of these animals have been excavated which show signs of having been conducted with special rituals. Since the bull is, very visibly, the epitome of power and strength, it is scarcely surprising that it should be selected as the sacred incarnation of those qualities. If a mortal man were able to vanquish such an animal, then surely he would have proved himself to be even stronger and more powerful, and thus himself a god. It is likely that such kingly bull-fights took place regularly, and that as time passed their form became increasingly ritualised, their progress stage-managed and their outcome predetermined and assured. In time, the god-king and the sacred animal became closely identified. The king gave himself such titles as 'The Strong Bull' and his artists, when carving reliefs in his praise, represented him as a bull, tossing his enemies in the air and overturning their fortified enclosures.

During the fourth millennium BC, these chiefdoms coalesced into two kingdoms, one around the lower and one the upper part of the river, and in about 3000 BC the two were brought together to form the first single-nation state the world had seen. A new capital for this united kingdom was founded, which was later called Memphis, just above the apex of the delta. The bull-god worshipped there was named Apis and it became one of the most important deities in all Egypt.

By this time, writing had developed and the abundance of Egyptian texts that survive from the following millennia reveal a detailed picture of how the people of Memphis regarded their gods. The creator of all was a god in human form they called Ptah, who conceived the world of both gods and men. Being a creator himself, he was a patron of those human creators, the artists and craftsmen. And he was incarnated on earth in the body of one special individual bull that was named Apis.

Only one Apis bull could reign on earth at a time. When he died, the priests searched the country for his successor. The new sacred calf could be recognised by very specific patterns. He was black, with a white inverted triangle on his brow. He had a pale patch shaped like a vulture with outstretched wings across his shoulders, a mark like a crescent moon on his flanks, and another the shape of a falcon which clasped his abdomen. His tail had double hairs on it; and beneath his tongue was a black mark shaped like a scarab beetle.

That a bull calf could be discovered that uniquely matched such an exacting description, even allowing for considerable imagination on the part of the priests who were charged with identifying it, tells

Capturing a bull A relief from the tomb in Saqqara of Mereruka, a nobleman who died about 2300 BC.

us something about the nature of Egyptian cattle of that period. One of the characteristics of fully domesticated mammals – whether cattle or goats, horses or rabbits – is that they frequently become piebald or vividly patterned in some way. In the wild any young born with such conspicuous markings would be obvious targets for predators and so tend to be removed from the breeding stock. In domestication, conversely, such a striking appearance might well catch the eye and the fancy of a herdsman. It made an animal conveniently easy to see and recognise at a distance. So such individuals not only survived but became specially selected and favoured, and bold variable patterning spread through the herds. Clearly this process was already taking place among Egyptian cattle.

The announcement by the priests that a young Apis bull had been discovered was a cause for national rejoicing. Children born on that auspicious day might be given the name 'Apis-is-Found'. The young animal was brought in procession to Memphis. On his way, he paid a visit to the temple sacred to Hapy, a fellow god directly associated with the annual flood of the Nile, and there he stayed while lavish preparations were made in Memphis to receive him. At this time, women alone were allowed to visit him. As they approached him, they stripped themselves naked and performed rites to ensure their ability to bear children. After several days, he was ferried by the priests down river to Memphis. There he was installed in the temple of Ptah in special quarters on its southern side, and there he spent the rest of his earthly life.

From now on he was regularly anointed by his priests with perfumed unguents and fed on special foods. During the regular fertility ceremonies, cows were brought to him so that he might mate with them as part of the sacred rites. On the anniversary of the king's accession, he accompanied the king in a procession around the boundaries of the state, so renewing the fertility of the fields. On festival days, he was brought from his stall to parade in front of the people with a garland around his neck and golden regalia fixed between his horns. Then his devotees might consult him. They did so by writing a statement, first in a positive and then a negative form on pieces of pottery, and placing one on each side of his route. They then deduced his answer by observing which he swayed towards as he walked.

When after twenty years or so, Apis died, his name was changed, to 'Osiris-Apis' so linking him with the god of the underworld, and his body was mummified with all the care and complex rituals that were accorded to a Pharaoh. The purpose of mummification, which the Egyptians had brought to a high degree of perfection, was to preserve the body as completely as possible so that the spirit might

The bull god of Egypt
In his representations, the diagnostic marks of the Apis bull are made unequivocally clear, and he is frequently shown with the regalia of the golden sun disc and cobra fixed between his horns. The bronze figure (*above*) dates from the third century BC, and comes from Memphis; the painting (*right*) was part of the decoration of the coffin of a priest who died in the tenth century BC.

An embalming table for a god
The temple of Ptah, where the Apis bull was kept, is now totally ruined; but two alabaster embalming tables dating from the sixth century BC still remain. The huge carcass of the bull god was placed with its head at the end with the head of the lion carved on the side, and lay there for sixty-six days while the priests completed the sacred rites.

return to it in the next world. The ritual required seventy days for its completion. Throughout that time, the priests undertaking it were in mourning, lamenting and rending their clothes, scattering dust in their hair and eating only a restricted diet of bread, vegetables and water.

The process began with the removal of the brain through a hole cut in the bridge of the nose. Then the stomach and intestines were taken out through an incision in the animal's flank. None of the entrails, however, were thrown away. All would be needed in the after-life and they were carefully washed, wrapped and placed in four special jars. Then the body was taken to the embalming table.

One of these tables, dating from the sixth century BC still lies today among a jumble of fallen columns and limestone blocks in a field outside the village that now stands beside the site of the temple. It is eighteen feet long and nine feet wide and made of the most beautiful veined alabaster. On each side of it is carved an immensely elongated figure of a lion, the guardian of the dead. Its surface slopes down gently towards a runnel at one end beneath which was found a huge alabaster bowl. For thirty-five days the body of Osiris-Apis lay upon it, packed inside and out with natron, an evaporite mineral consisting of a combination of sodium carbonate and bicarbonate and other salts that occurs naturally around the edges of salt lakes in Lower Egypt. This sterilised the flesh and absorbed fluids from it so that the body dried out. After thirty-five days, the corpse was bathed with embalming fluids – milk and wine scented with juniper oil, beeswax and spices. As these fluids ran down the table, they emptied through the runnel and were collected in the trough beneath. Having been in contact with the body of a god, they were magically very powerful and had to be disposed of with care. It is possible that they were sold by the priests to the devout. Then the whole body was covered with molten resin.

Now the bandaging could begin. This took about thirty days, every action being accompanied by magical words and ritual gestures. On the 67th day, the mummy, garlanded and decorated, was placed on a sledge and pulled in procession, accompanied by the priests, a military guard and a representative of the Pharaoh, to a ceremonial barge that awaited it on the shores of the lake separating Memphis the city of the living, from Saqqara, the city of the dead.

On the far bank, the procession made its way along a ceremonial avenue lined by 134 stone sphinxes that led up to the temple dedicated to Osiris-Apis. This building has now totally disappeared but hewn in the rock beneath it was a catacomb in which all the bulls were finally laid to rest. Its discovery, in 1851, was one of the earliest sensations of Egyptian archaeology and it remains one of the most astounding. It is known today as the Serapeum, from Serapis, the Latinised version of the name Osiris-Apis.

Preparations for the reception of the bull mummy had been going on for a long time, often even before the bull had actually died. A vault had been cut in the rock on one side of the spacious central corridor of the catacomb and a sarcophagus placed there ready for the mummy. In the early years of the cult, this was made of wood, but by the seventh century BC the Pharaoh himself provided one made of granite. It was cut from quarries 400 miles up-river beside the First Cataract of the Nile, ferried down by barge and then dragged by the masons across the desert to the temple and finally down into its destined vault. It weighed about 65 tons.

The floor of each vault lies fifteen feet below the level of the corridor. To lower the sarcophagus into place, the vault was first filled with desert sand so fine and so dry that it flows almost like a liquid. The sarcophagus was hauled on to the sand filling the vault and then lowered by removing the sand from the sides as it was expelled laterally by the great weight of the stone.

The catacomb of the bull gods
The deep vaults, each of which contained an Apis bull, have been cut from the rock on either side of the long corridor of the Serapeum. The immense granite sarcophagi still stand in them, with their lids on the level of the corridor floor; but all were robbed in ancient times.

When the bull mummy itself arrived, it was carried along the corridor, into the vault and then down into the sarcophagus. The alabaster jars containing the viscera and many offerings to serve Osiris-Apis in the afterlife were placed around it. Then the immense granite lid, itself weighing many tons, was hauled over the top of the sarcophagus, sealing the body of the animal god within to await eternity.

But it did not remain so for long. The Serapeum was in use for over a thousand years. When it was rediscovered in the nineteenth century, all but one of the bull sarcophagi had been broken into and plundered, the lids pulled askew, their corners crudely battered off to allow a child to slip inside and remove anything of value. Only one burial remained intact. In it, beside the mummy, the excavators found a superb statue four feet high of an Apis bull, an offering made from solid gold.

The bull, by virtue of its long relationship with the founding dynasty of Egyptian kings, was one of the greatest of the state gods, but it was not by any means the only animal to which the Egyptians attributed supernatural powers and divinity. Indeed, their attitudes towards animals, moulded by 3,000 years of theological elaboration, ultimately became so full of complexities, conflations and contradictions that almost any generalisation about them can be contradicted by one example or another, any rationalisation about their origins or responsibilities confounded by an illogical conclusion.

The crocodile, for example, was understandably seen as a force for evil. It habitually lay half-hidden in the water. It attacked women and children as they crossed a stream and killed cattle which came down to drink. On the other hand, it was a source of good, for when the annual floods of the Nile receded each October, mud-banks laden with rich silt were the first signs of the reappearing land, and on them, in a proprietorial way, lay the crocodiles. So the crocodile god, Sobek, was feared as an associate of Seth the god of evil, together with the hippopotamus and snakes, but at the same time worshipped as Lord-of-the-Island-in-the-River, who brought back land to the earth. In one temple, crocodiles were kept and bred in two stone-lined pits fitted with stone covers that could be rolled back on small bronze wheels so that the animals could be fed. But in the river outside, they were hunted mercilessly with spears.

A mummified cat
In later periods of Egyptian history, all animals were believed to embody some part of the spirit of the patron god whom they represented on earth. They were all, therefore, suitable subjects for mummification. This cat was embalmed about 30 BC.

The cat was also worshipped. It lived wild in the valley but had, probably of its own accord, taken up residence in the houses of the people, certainly by 1600 BC. Once there, it was welcomed as an ally that helped to keep down the number of rats and mice. Whether or not it was domesticated in any real sense is difficult to say, for until recent times man has had little influence over its breeding; cats living with people have hardly differed anatomically from those living completely wild lives. The Egyptians seem to have had a great affection for them: the death of one was the cause for a whole family to go into mourning. It is the more surprising, therefore, that this amiable if aloof member of the household should have been regarded as an associate of Pasht, the goddess of war, whose name is probably the cause of our calling a cat 'puss'. The reason may lie in the fact that Pasht herself was believed to be incarnated as a lioness, a suitably ferocious creature; and the Egyptians doubtless saw a cat, quite correctly, as a diminutive version of her.

The baboon was also kept as a pet. When, in later times, it had virtually disappeared from Lower Egypt, animals were sent down specially from the upper parts of the valley where it was still common. In paintings and reliefs it appears frequently – stealing figs and grapes, sitting beside ladies at their toilet, even beating drums and dancing. Yet it also has a wise air and is obviously intelligent. So in one form it became one of the lesser gods, Baba (from which comes its English name), and given the engaging title of 'He-of-the-Red-Ears-and-Violet-

Bottom'; and in another, it is one of the most powerful spirits in the whole Egyptian pantheon, Thoth the god of wisdom and learning. By extension, Thoth became the patron god of writing and therefore of scribes; and so was connected with that marker of time, the moon. In consequence, he is often represented as a baboon with a moon disc on his brow.

But gods might be incarnated in more than one form. The handsome black and white ibis, which we still call the sacred ibis, is now common only south of the Sahara, but in ancient times it was abundant in Egypt. It feeds by gravely probing the muddy fields with its long curved bill, and this habit was interpreted by the people as a continuing search for truth. Furthermore, the bird's shining white colour linked it with the moon. So it, too, was believed to be an incarnation of Thoth.

The falcon, perhaps, had one of the clearest and least ambiguous supernatural identities. In the sky, ceaselessly scanning the earth beneath, on occasion flying so high that it almost disappeared, it became identified with the sun. So it was seen as Lord of the Sky and was depicted bearing the sun disc on its round head, with the sun's rays flashing from its wings representing the sky. Such a powerful creature was, therefore, like the bull, appropriated by the Pharaohs. So Horus, the falcon god, is represented with a sun-disc or a crown on his head; and the statue of a Pharaoh often has a falcon perched on the back of his neck enfolding his head-dress with its outstretched wings.

In early times, it was believed that only one individual animal, among the multitude of the same species that existed, was the earthly incarnation of a god, as Apis was among ordinary bulls. So while that animal was kept in the temple and treated with the greatest reverence, other animals like it in the world outside could quite permissibly be maltreated, killed or even eaten. But as the centuries passed, so people began to believe that the spirit of a god was not restricted to one animal, but that all individuals of that species contained a fragment of divinity and were to some degree sacred. They were all, therefore, like the Apis bull, suitable subjects for mummification. Accordingly, arranging for *any* animal to be embalmed was a way of gaining favour with its patron god.

By the sixth century BC the Egyptian people had taken to this practice on a scale that beggars belief. In Saqqara, not far from the Serapeum, in the wasteland of shifting sands where once stood a group of splendid temples, a small steep shaft has been cut into the sandstone, which leads down to a small chapel. Within this there is a throne on which, almost certainly, there once stood a statue of Thoth in his incarnation as a baboon. Excavated in the wall to one side of it,

The sacred ibis
This species of ibis is still called 'sacred', even though it has now been driven from the land where it was once revered as a god. These are feeding in a swamp in east Africa.

invisible to anyone approaching it directly along the shaft, is another smaller cell, just big enough for a man to crouch in, which can be entered from the back. Here, it seems, a priest concealed himself and answered questions put to the god by pilgrims, perhaps by speaking, perhaps by barking like a baboon.

Beyond, the passage tunnels into the rock. Immediately inside, niches have been cut into the walls. These once held wooden boxes in each of which lay the embalmed body of a baboon, packed around with powdered gypsum to protect it from damage by vibration. Some had inscriptions on the stone slab that blocked off the niche, giving the name of the baboon that lay within, its age, the date on which it was installed in the temple and the date of its death, together with prayers for its spirit. Some of these animals seem to have lived for several years in the temple as incarnations of the god. But well over 400 baboons were deposited there in the 500 years that this catacomb was in use, so many of them must have been brought to the temple specially for mummification. But greater astonishments lie ahead.

After a few yards more, the narrow passage opens out into a relatively spacious corridor seven or eight feet high. It is full of elongated pottery jars. They are between eighteen inches and two feet long, with lids secured in place with mud or a sandy plaster. They lie on their sides in stacks like wine bottles. Chambers cut in the rock on either side of the corridor are piled to within a few inches of the ceiling with them, twenty or thirty rows deep. More are stacked waist high on both sides of the corridor so that in places there is only just room for one person to walk between them. Some of the lids of the pots have fallen off to reveal the mummified body of a bird. The bigger pots contain ibis; the smaller contain falcons. Many are most elegantly wrapped in linen bandages laid in careful diagonals with the features of the bird's head painted on them in black ink, and a false beak made of wadding carefully attached so that the species is immediately recognisable. Some have been given much more summary treatment and are enclosed only in a bag of coarse sacking, soaked in some tarry substance and criss-crossed with string. Among them are one or two very small pots. They contain the embalmed bodies of shrews, presumably so that the spirit falcons could have spirit animals on which to prey.

Within twenty yards, the passage divides. More chambers lie down either side of each branch. Again the passage forks, and yet again, until eventually a branch ends in a blank stone wall. A small hole in the top corner explains why it goes no further. Peering through with a torch you can see beyond yet another chamber stacked to the ceiling with pots. If the workmen had tunnelled any further, they would have broken into another corridor coming from the opposite direction.

This labyrinth of galleries, which was discovered less than 20 years ago, has not yet been fully explored. It may never be. In places, rock falls warn of the danger that might come from the first footsteps to echo in these extraordinary corridors for two thousand years. A conservative estimate, however, is that eight hundred thousand falcons and four million ibis are lying here, each in its own pottery sarcophagus. The true number may well be twice that.

The birds were offerings made by devotees of Horus and Thoth, who visited the temple that once stood on the ground above and wished to gain favour with the gods by presenting an embalmed bird which might contain a tiny particle of the divine spirit. But where could such vast numbers of dead birds be found? Falcons and sacred

The falcon god
A gigantic granite image of Horus (*above*), wearing the double crown of Upper and Lower Egypt stands beside the main entrance to the god's temple at Edfu on the Upper Nile. At Saqqara in Lower Egypt, the birds were mummified, sealed in small pottery sarcophagi (*above right*) and stacked by the thousand in underground galleries.

The baboon god
Thoth, being the god of wisdom, was frequently consulted as an oracle, questions being addressed to its image and the answers being interpreted by a priest. This statue (*right*) stands at the entrance to the baboon catacombs at Hermopolis.

ibis were certainly kept in the temple enclosure, but natural deaths among them could not conceivably meet such a huge demand. Furthermore, killing these sacred birds was now regarded as a crime so serious that it was punishable by death. The priests undoubtedly found a solution to the problem. Close beside the site of the temple lies the dried-out bed of a lake. Here they must have raised the birds that later they killed, embalmed and sold, ready-wrapped, to the devout. Clearly their interpretation of religious prohibitions was somewhat elastic. Nor were they over-particular ornithologically. Among the falcon mummies are the bodies of quite different birds, and the wrappings of some contain nothing more than a miscellaneous collection of bird bones roughly bundled together.

The Saqqara galleries are by no means unique. At Hermopolis, one of the most important cult centres of Thoth, there were more catacombs full of ibis and baboon that have long been known. At Kom Ombo, on the banks of the Upper Nile crocodiles were mummified. During the last century, three hundred thousand embalmed cats were discovered at Beni Hasan in Middle Egypt. Twenty-four tons of them were dumped on a cargo ship and sent to Liverpool to be sold for fertiliser, an enterprise that – perhaps fortunately for Egyptology – proved to be a commercial failure.

THE ANIMAL-HEADED GODS

Egyptian artists, when portraying animal gods, often represented them with human bodies, so making clear their divine nature and the difference between them and earthly animals. Horus, the god of the sky who was closely connected with the Pharaoh, had the head of a falcon (*2*). Thoth, the god of wisdom, might take the form of a baboon or an ibis (*3*), here with a sun-disc on its head. Anubis, the jackal god, (*4*) deity of the city of the dead, presided over the rituals of mummification (*5*).

1

2

3

The Book of the Dead is a collection of religious texts and spells placed in the tomb of a human mummy to allow the spirit of the deceased to leave the tomb when necessary. An illustration from it (*1*) shows some of these gods caring for a human spirit. The dead man, on the left, is led by Anubis to the ceremony in which his heart will be weighed. As this is done Anubis, shown a second time, carefully checks the reading on the scales while on the other side Ammit, a meat-eating monster waits to consume the heart if it is found to be sinful. Thoth writes down the result and the dead man, having passed the test, is introduced by Horus to Osiris, ruler of the underworld. Above, the dead man, his arms raised in worship, sits before fourteen gods and goddesses who have been witness to the weighing.

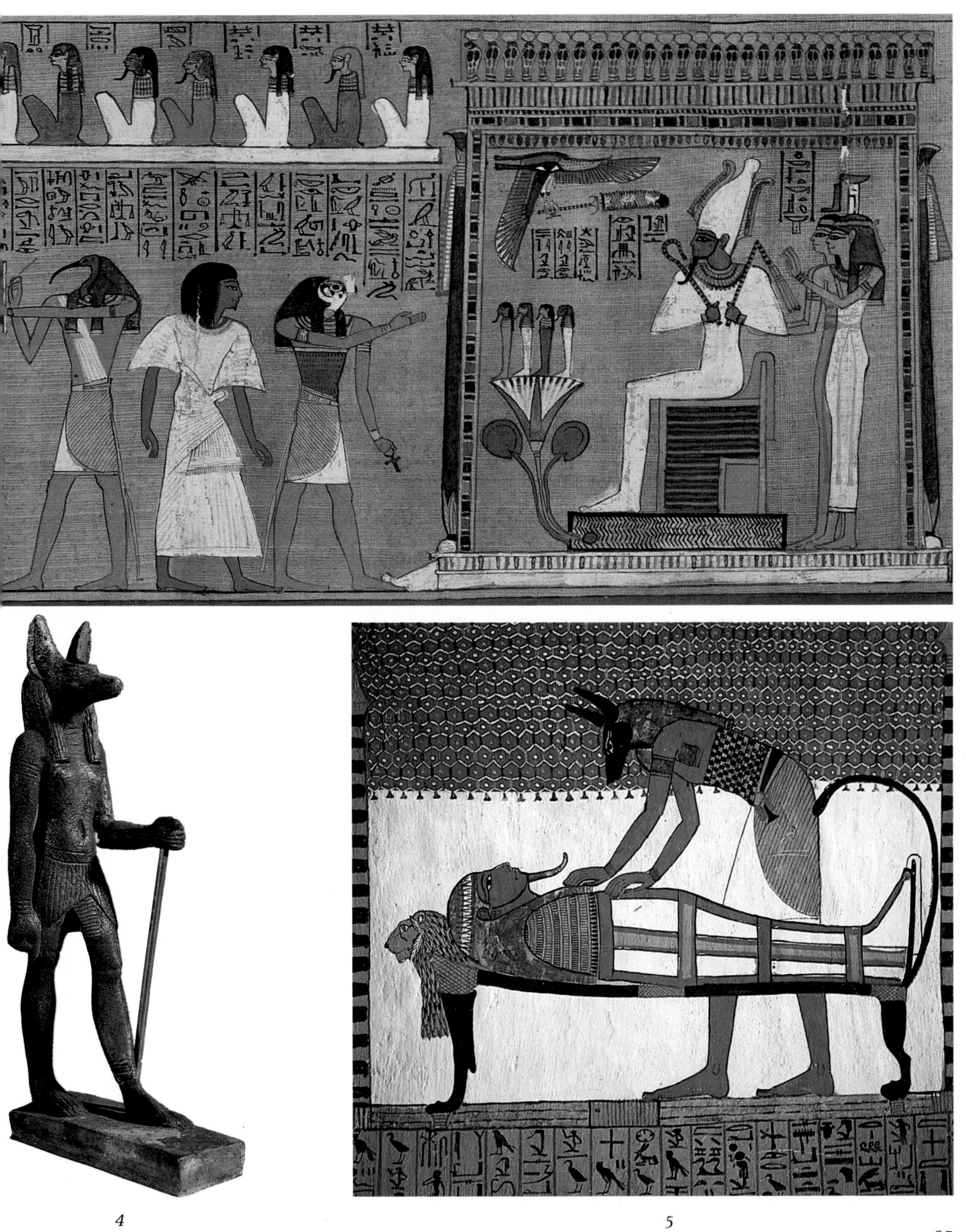

4

5

Taweret (*2*), the swollen-bellied hippopotamus god, is one of the few whose human body is clearly female. She was a protector of pregnant women, and assisted in child-birth. An illustration from the Book of the Dead (*1*) shows a woman praying to her. Sekhmet (*3 and 4*) who has the head of a lioness, was the patron deity of physicians and also the consort of Ptah, the creator god. Sobek (*5 and 6*), the crocodile god, another deity involved with creation, brought back the land after the annual flood.

1

3

4

5

2

7

6

8

Khnum (7), was also connected with the powers of creation. He had the head of a ram, whose appetite for mating was clearly well-recognised. The cult of Pasht, the cat-headed god of war, was centred around the city of Bubasti in the Delta where this image (8) was found. She holds a musical instrument, a sistrum, in her hands, and cats sit at her feet.

A muster of animals

These paintings come from the tomb of a high official from Thebes and date from the fourteenth century BC. They show the domestic animals belonging to the estates of the Temple of Amun being brought in for their annual census.

The cattle, which are noticeably different from one another in colour and pattern, are being urged on by a boy. The geese, too, are clearly domesticated stock, for they are also much more varied in colour and pattern than they would be if they were truly wild.

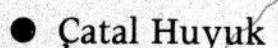

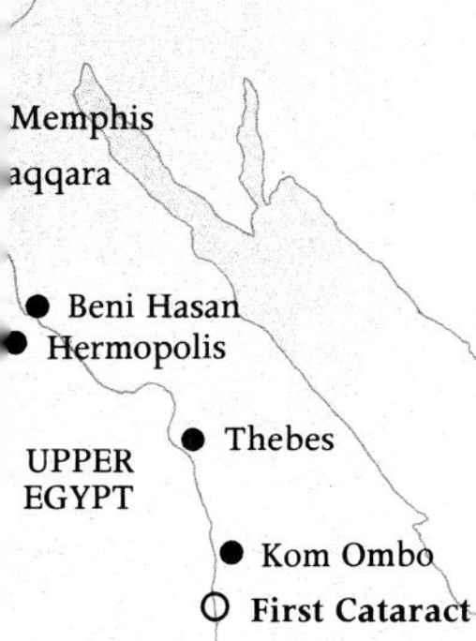

While the Egyptians saw divinity in all the creatures around them, they had no inhibitions in handling and exploiting them. Once again, the most vivid evidence comes from wall paintings and reliefs. These are in the tombs of courtiers, priests, soldiers and officials, not in those of the Pharaohs where the decorations are, for religious reasons, restricted to ritual scenes and representations of the underworld. Nobles, however, regularly included in their tomb decorations pictures of their earthly life among the ritual funerary scenes, for they believed that scenes painted on the walls of the tomb in which their body would ultimately lie would be recreated in the afterlife. So naturally enough, they chose as subjects those events and activities they had most enjoyed. The artists they employed portrayed nature with such accuracy and obvious knowledge that many species of animals and plants are easily recognised – hoopoes, kingfishers and red-breasted geese, papyrus, date palms and lotus.

The people no longer needed to hunt in order to feed themselves. But hippopotamus had to be cleared from the rivers and canals for the safety of the boat traffic, so hunting parties went out to pursue the animals – and clearly did so with great enthusiasm. The wealthy went out into the desert on hunting trips and brought back, hobbled and haltered, ostriches, giraffe and gazelle.Captive oryx may have been kept like domestic animals, for some reliefs show their calves being born with the help of human attendants.

The skills of animal husbandry are shown in detail. Geese and cranes are force-fed, as were hyaenas, if the reliefs are to be believed. Cows are mounted by bulls, give birth and are milked. Herds are led across rivers by a man carrying a calf on his shoulders which looks back piteously towards its mother swimming behind. Bulls are lassoed, thrown on their backs and await slaughter, with bowls beside them in which to catch their blood.

One of the most splendid of all these noblemen's tombs belonged to Menna, a high official who served King Thutmoses in the fourteenth century BC. On one wall he is portrayed on a wildfowling expedition in the papyrus swamps. A trained cat has been sent into the reeds to scare the birds into the air, while Menna hurls throwing sticks at them. Several tumble from the sky. Menna himself holds some by their legs. One of his servants, depicted as a very small figure as befitted his relative unimportance, holds others by the wings. On

another wall, Menna throws a harpoon at a shoal of fish. The delight that he and his family – and indeed the gifted painter who portrayed them – took in the natural world is very vivid. His little daughter kneels beside a pool and tenderly plucks a blue lotus bud; butterflies flutter through the reeds; a small mongoose clambers up a papyrus stem which bends under its weight; and, in the water beneath, a crocodile captures a fish between its jaws.

Menna's job was to inspect the estates, supervise the harvest and assess it for tax purposes. This, too, is recorded in detail. Across the fields cattle draw a plough which, in design, is identical with those that can be seen in use today. The grain is scattered, sprouts and is measured as it grows. Workers gather the corn heads with flint sickles, and cattle tread the harvest on the threshing floor. Other men winnow and carry away the grain in huge baskets for storage.

The fertility of Egypt must have seemed to its people to be eternal. Its richness was emphasised by the contrast with the desert that hemmed in the valley on both sides. Within its boundaries nature could hardly have been more mysteriously beneficent. In spite of the fact that virtually no rain whatever fell on Lower Egypt, the great river rose unfailingly every year. As it did so it changed colour. In June it was green; by August it had turned brown; and come September, it was so swollen that it burst its banks and flooded the country. For six to eight weeks its waters lay on the land. Then, every year, the flood retreated leaving behind a fresh deposit of black fertile mud so that once again the crops would sprout in abundance.

The Egyptians knew nothing of the highlands of Ethiopia where the rain falls that feeds the Nile. They believed that Hapy, the god of the flood, brought forth the annual surge of water from subterranean basins somewhere in the region of the First Cataract where the river rushes through a wide band of granite rocks. They thought that the fertility of their animals was maintained by the power of Apis the bull god, that the sun shone each day on the sprouting corn under the influence of Horus the falcon. So, over three thousand years, the people of the valley grew richer and richer, their cities and temples more and more glorious and extravagant, until, by the first century AD, seven million people were living there. They produced so much grain that they became even richer by exporting their surplus to less favoured lands. Egypt was, indeed, blessed by its gods.

Hunting in the marshes of the Nile
Birds rise in panic as a nobleman approaches in his boat. In one hand he holds a throwing stick, in the other three herons. His cat has already caught three birds. His wife stands behind, and his daughter squats to pluck lotus from the fish-filled waters.

Life on an Egyptian farm
Sen-nedjem was a nobleman who lived in Thebes during the twelfth century BC. One wall of his tomb is occupied by a picture of him and his wife at work in the fields. Irrigation channels full of water surround the estate. Date palms loaded with fruit, and bushes thick with flowers grow around its margin. In one part of the fields, Sen-nedjem and his wife gather flax. In an adjoining plot, he drives a pair of dappled cattle yoked to a plough while his wife walks behind him, sowing the grain. In the scene above, he cuts the ripe corn heads with a sickle. Such pictures were, in themselves, allegories of the resurrection. Just as the crops are cut down, resown and then spring again, so a man will find life after death.

The unchanging olive harvest
Olives are grown from clones and so the fruit of many ancient trees may be exactly the same in size and taste as that collected by the ancient Greeks. The methods of gathering the olives are also unchanged. A vase painting made around 520 BC (*top*) shows men with sticks knocking down the fruit on to sheets spread beneath. People in Greece still gather them in identical fashion (*above*).

THE KINGDOMS OF THE BULL GOD

The islands lying north of the Egyptian coast in the middle of the Sea – Cyprus, Crete and the smaller fragments of land that make up the Cyclades – were not so well-favoured. They had no great rivers to replenish their fertility. Their land was rocky and their soil was thin. Nonetheless, by 6000 BC people from the mainland coasts had reached them and settled there together with their domesticated plants and animals – wheat, cattle, sheep and goats. And in the forests of oak, cypress and fir, they found a tree that was to become a source of great wealth to them, the olive.

The light, well-drained soil of the islands suited the olive, and its leathery leaves enabled it to survive the hot, rainless summers. Its fruit, consisting of a stone surrounded by a coat of oily flesh, was edible; but the crop produced by the wild trees was very variable, both in quantity and succulence. However, a tree could be propagated not only from seed, but also from small knobs that develop on the side of its trunk close to the ground. If these are cut off, they readily take root. The offspring produced in this way are clones, genetically identical to their parent. So if an exceptionally prolific tree with particularly oily fruit was discovered growing in the forest, it could be replicated over and over again. Furthermore olives are exceptionally long-lived. Some may survive, hollow-trunked and gnarled, but still bearing fruit, for as much as 1,000 or even 1,500 years.

So slowly, over the centuries, the people assembled whole plantations derived from carefully selected individuals. When and where this habit started is not certain. The wild tree grew, and still grows, both on the islands and in most countries all around the Sea in Africa as well as Europe. Increasing oiliness and abundance of crop are very difficult to detect archaeologically, but orchards of clones will produce crops with stones that are very similar in shape and size, and from this clue it seems likely that domestication began somewhere in the eastern Mediterranean around 6000 BC. The earliest sign of the use of the oil comes from the islands and was found in a grave excavated on the Cycladic island of Naxos and dated to the third millennium BC. It is a small jug in which archaeologists managed to detect minute traces of oil.

We do know how the fruit was gathered. A Greek vase, albeit from a much later period, shows men beating the branches of an olive tree with sticks and knocking down the fruit on to sheets spread beneath. People do exactly the same thing today. The ripe fruits were then drenched with hot water and pressed. The fluid so produced was left to settle in vats and the oil skimmed from the top.

The clear, fragrant olive oil was used in three ways. It provided the main source of light, being burned from a wick in small lamps of pottery or stone, two of which were found with the tiny oil jug in

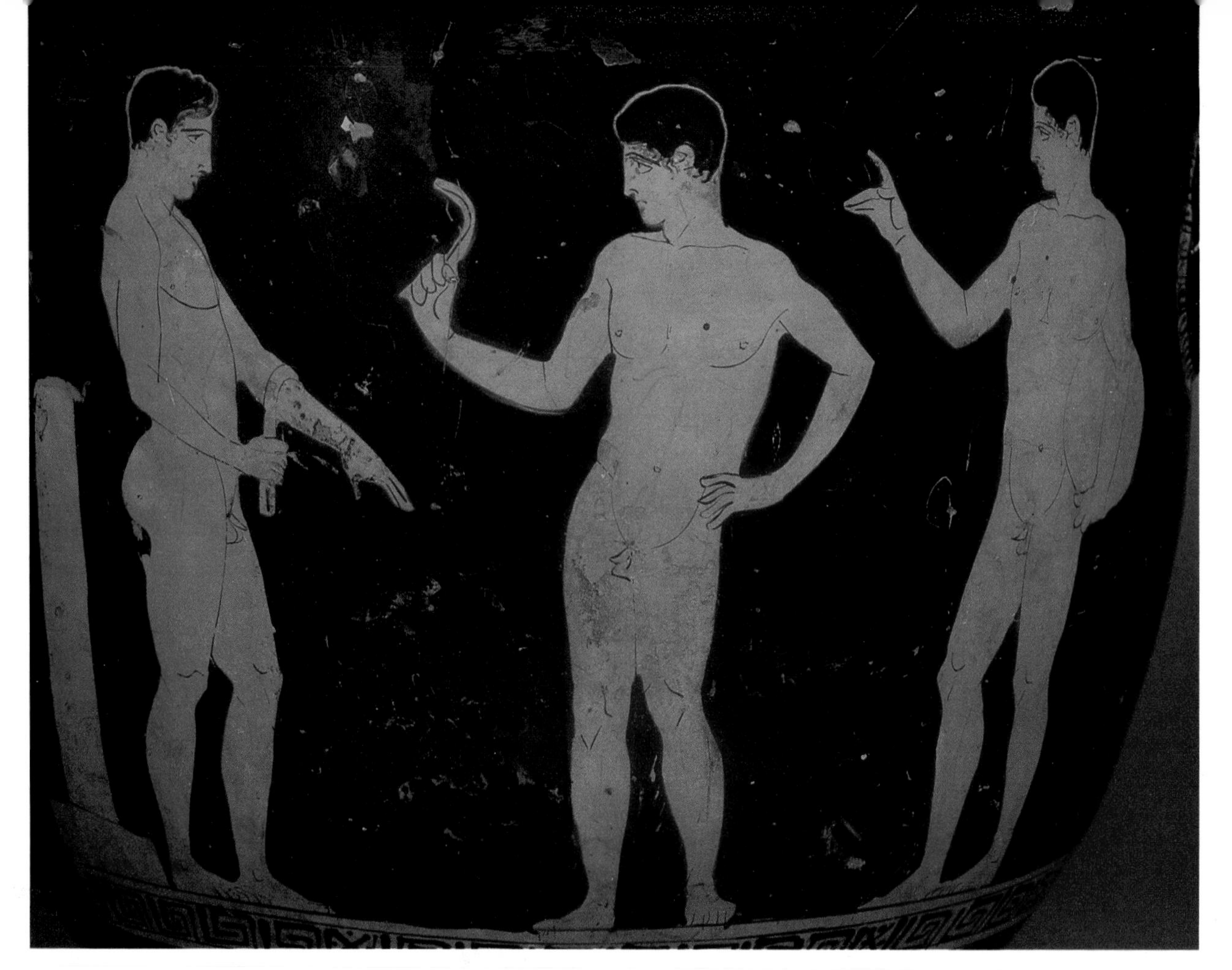

Oil for the body

After strenuous exercise, both Greeks and Romans cleaned their skins of dirt and sweat by anointing themselves with olive oil, and then scraping it off with a curved blade of wood or bronze, the strigil. These athletes at their toilet were painted on a Greek vase in the 5th century BC.

the Naxos grave. It was also important as a cosmetic. After a day's work in the hot dusty fields, people anointed themselves with it, often mixing it with sweet-smelling spices. It restored suppleness to the skin, soothed abrasions and the burning and drying caused by sun and wind, and also, in the absence of soap, carried away the dirt. The surplus oil was then scraped off with a strigil, a curved blade of wood or, in later times, of bronze. Thirdly, it was used in the kitchen, as it is today, as a food in itself and as a medium in which to cook other foods.

That early Naxos grave also showed that the people had discovered another valuable plant in their forests, the grape vine, for it contained a small cup, shaped like a sauce-boat, of a kind that was used for wine drinking. The wild grape vine is a creeper that twines around forest trees and it too was widely distributed in the lands of the eastern Mediterranean. Indeed, its range was even greater than the olive, because it is better able to withstand the cold, and therefore it grew somewhat further north. It, too, could be propagated as a clone, by taking cuttings, and it seems that the people understood at a very early date the techniques of grafting. This enabled them to take, for example, a shoot from a plant that bore abundant sweet fruit and attach it to a root stock that was particularly vigorous or resistant to disease.

Domestication in the vine can be recognised because in the course of that process, the pips of the grapes became bigger and more elongated. From such evidence, it seems that the vine came into cultivation at much the same time as the olive, and in much the same area. And the people discovered very soon that grapes were not only valuable as fresh fruit. Their juice, when fermented, made the most delectable drink.

The sea, too, yielded abundant food, and the Mediterranean was more welcoming than most. Tides and strong currents which give seamen so many problems on oceanic coasts, are largely absent; the weather is generally calm from April to October; and the islands themselves, in the eastern part of the Sea, are so thickly scattered that land, on a clear day, need never be out of sight. So the people, who must have been very competent seamen to have reached the islands in the first place, never deserted the sea and were expert in exploiting its riches. They gathered molluscs not only for food but to use in the dyeing of cloth. They collected coral and pearls for use in jewellery. They trapped octopus in pottery traps, and gathered sponges. And they fished intensively with nets and with hooks.

Harvesting the sea
The inhabitants of the sea were a favourite subject for Roman mosaics. This one, from Sousse in Tunisia, also shows some of the methods used to catch them – cast nets, seine nets, rod and line, and lobster pots.

The cult of the bull was also practiced in these islands too. In Cyprus, a mere forty-four miles away from the Turkish coast, pottery images of bulls appear around 2700 BC. As the centuries passed they were produced in great numbers and considerable variety. Some are free-standing figurines with heads lowered and horns pointed belligerently forward. Others appear with deer and human figures around the necks of pottery vessels. In one remarkable instance, they are placed within a bowl in an elaborate group that may represent either a village scene or a religious ritual in a temple. Four bulls stand in stalls, human figures sit or stand around them, and modelled on the side of the bowl are three poles each surmounted by a bull's head. Small terracotta models of what are undoubtedly shrines have also been discovered which again show poles topped with bulls' heads. The sites of actual sanctuaries have been excavated and yielded huge numbers of pottery figurines, including a striking figure of a bull being led by two men as if to a sacrifice, and human figures wearing bull masks.

Bull worship in Cyprus
The potters of Cyprus, working around 2000 BC, left intriguing evidence of their daily lives. They made models of the shrines at which they worshipped. One (*left*) shows three poles or pillars surmounted by bull's heads and a human figure standing in front, perhaps pouring a libation into a large vessel. A remarkable bowl from the same period (*far left*) shows a gathering of people, some standing, some sitting, while cattle (*below*) wait in a pen.

In Malta, at the other end of the Mediterranean, the people built their temples from huge stone blocks. For the most part they are devoid of decoration. Where it does appear, it is for the most part, abstract geometric pattern, but among the very few naturalistic designs there are, once again, reliefs of bulls. And the memory of the supernatural power of bulls lingers in the island even today, for the people still mount bull skulls, horns or modelled heads high on the walls of their houses to protect themselves from bad luck.

Farther west still, in Sardinia, rock cut tombs dating again from the third millennium BC are decorated with the heads of bulls. The bull cult about which we have most evidence, however, is that which was practiced in Crete.

By 2000 BC, the Cretans had begun to establish colonies in other islands which eventually enabled them to dominate the islands of the eastern Mediterranean. Their ships travelled south to Egypt taking olive oil, wine and cypress wood for trade. They went north, too, to collect obsidian, a volcanic glass with which they made tools, and there they established new settlements. As they prospered, they began to build large towns. Ancient accounts say there were as many as 100 on Crete alone. Of the four that so far have been excavated there, the largest is Knossos, a few miles from the north coast. At its centre stood a great palace, roughly rectangular in plan, with sides about 150 yards long. Some of its buildings were two storeys high, supported on stout columns of cypress wood. It had elaborate drainage and sewage systems, and in its basement lay vaults filled with rows of huge pottery jars, some over five feet high. In all, over four hundred of them have been found. Some had contained wine, others olive oil.

The treasury of King Minos
Huge storage jars lined the vaults of the palace at Knossos. Some were filled with olive oil, others with wine, riches that were probably brought to the palace as tribute.

This was the state's treasury, where the produce of the land was sent as tribute. It has been estimated that at least eighteen thousand gallons of oil was stored here, and that such a quantity implies the existence of plantations containing at least thirty-two thousand olive trees.

The palace shows no signs of battlements nor other defences. It was not a fortress, but a religious and economic centre, and it was here that the bull was worshipped. Bull images appear over and over again through the building. They were painted on pottery vessels and on the plastered walls. They were carved in marble and used to mark specially sacred places. Earthquakes, which in this volcanic region are not infrequent, were attributed to the bull lying within the land shaking himself. Tidal waves that accompany volcanic eruptions were also manifestations of his power. The god-king himself, whose name was Minos, was believed to have had a bull as an ancestor and was therefore half-bull himself and the incarnation of the bull spirit in human form. On ceremonial occasions, he and his queen, dressed as a bull and a cow, may have enacted fertility rituals similar to those that took place in the fields of that other bull-king, around Memphis in Egypt.

The Knossos palace, like the three others so far excavated in Crete, was built around a central courtyard. There, many archaeologists believe, the ceremony that was the main element in the bull cult was held. It was a blend of religious devotion, athletic prowess and great bravery, and of such an extraordinary character that were it not vividly depicted on wall paintings, pottery vessels and bronze statuettes, it would be difficult to believe it took place at all.

A drinking cup from the Palace
Among the abundant images of the bull found in Knossos is this superb drinking vessel. Carved from black steatite, its eyes are of rock crystal and the white line around its muzzle of inlaid shell. Its wooden, gilded horns were missing and have been restored.

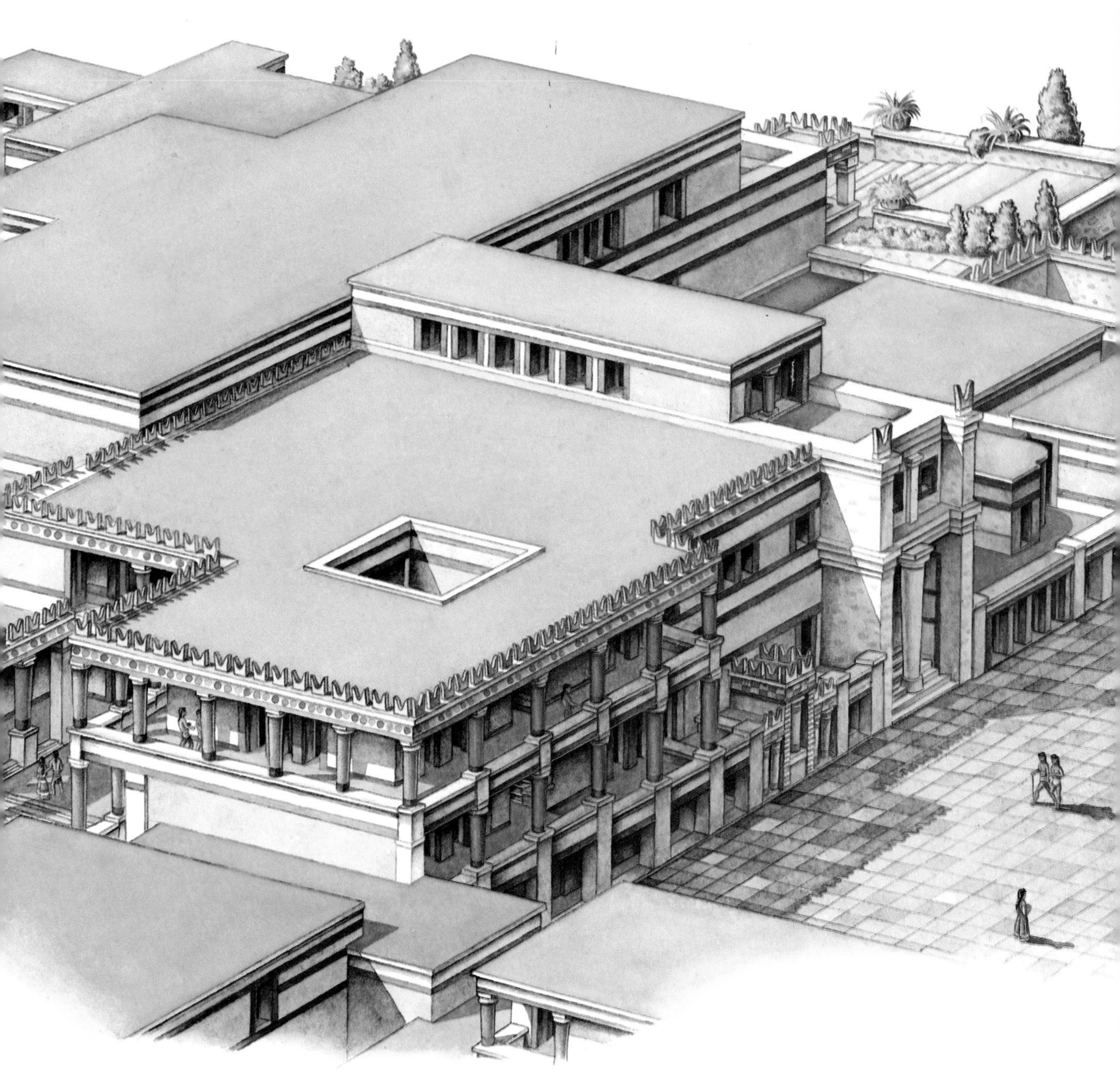

The palace of the bull king
The buildings that made up the palace and administrative centre of Knossos were grouped around a vast central courtyard where the bull vaulting ritual was probably held. Many of the roofs were edged with the two-pronged symbol of the bull horns. The colonnaded galleries, their wooden columns wider at the bottom than the top, were decorated with brightly coloured frescoes. This is a reconstruction of the range that stood on the western side of the courtyard and included the royal apartments and the throne room.

Bull capture
The golden cups of Vaphio illustrate the way that bulls were captured alive in the forest. A cow is used as a decoy (*top*). A net is spread between the trees in which the bull becomes entangled (*middle*), though in the process a man has been thrown to the ground. Finally (*below*) the captured bull is hobbled with a rope.

Bull leaping
A fresco from the wall of a room in the Palace at Knossos (*below*) shows in a single design three phases of the leap – the athlete seizing the horns of the charging bull, somersaulting over its back and landing on the ground behind it. Other evidence of this extraordinary practice comes from engraved seals and a bronze statuette (*above*).

Bulls, the descendants of the early domesticated herds that the first settlers from the mainland brought with them to the island, roamed wild in the Cretan forests, and there the people trapped them alive. Reliefs embossed on two superb golden cups discovered at Vaphio on the mainland, but of Cretan inspiration if not manufacture, make clear how they did it. Rope nets were stretched between trees and a bull enticed into them by the use of a decoy cow. Once captured, the bull was transported to the palace and there, on the appointed day, in front of the assembled people, it was released into the arena. An athlete stood facing it. As it charged, head down, he grasped it by its horns and allowed himself to be tossed upwards so that he somersaulted in the air and landed on his feet either on the bull's back or on the ground beyond it. Sometimes it seems a variant was performed in which the athlete side-stepped, grabbed just one horn of the animal as it passed, and swung himself across to grasp the other, and so vaulted over the bull's neck. Such feats were obviously extremely dangerous. Attendants, sometimes girls, stood by to steady the vaulters as they landed and, doubtless, to divert the bull when necessary; but, as designs on engraved seals record, sometimes the performer fell and was trampled beneath the bull's hooves.

Crete's domination of the eastern Mediterranean lasted until 1500 BC. Then, with extraordinary abruptness, its power disappeared. Until recently, many believed that this was caused by a catastrophic eruption of the island of Santorini, 120 miles away to the north where the Minoans had a colony. This eruption was certainly one of the biggest in historic times, and half the island was blown away. Some authorities suggested that ash fell so thickly on Crete that the crops were totally destroyed, that huge earthquakes toppled the palaces and that great waves, produced by the eruptions, swept inland and overwhelmed the towns. But no certain evidence of any of these phenomena on a scale big enough to cause such an effect has yet been found on Crete, and now careful dating has shown that the eruption took place fifty years or so before the collapse of the Minoan cities.

So the cause of the collapse of Cretan power is still uncertain.

The cultural leadership of the eastern Mediterranean now moved first to the Mycenaeans who lived on the great Greek island of the Peloponnese, and later to mainland Greece. The Greeks retained a memory of the one-time domination of Crete in their myths which told how, every nine years, they had been compelled to send a tribute of seven young men and seven young girls to Crete. There they were sacrificed to a monster called the Minotaur, half-man half-bull, that was imprisoned in a labyrinth at Knossos. They only gained their freedom when one of the Greek heroes, Theseus, sailed to Knossos, found his way into the heart of the labyrinth and there slew the monster.

While acknowledging that Zeus, the creator of all, had his origins in Crete, the Greeks saw their gods very differently from their Minoan predecessors. The bull does not appear among them. Neither is any of them animal-headed as those of Egypt had been. Instead they are all very human, both in form and character. They squabble and steal from one another; they tell lies, philander and become jealous. They even make love to mortals. Animals are demoted to become their attendants or convenient disguises in which to disappear during their adventures. Zeus turned himself into a swan in order to seduce the beautiful mortal, Leda. And to conceal Io, another of his loves, from his jealous wife, he simply turned her into a white heifer.

But on the eastern frontiers of the Greek states, in Asia Minor the fertility gods and the animal cults, that had been so important in Çatal Huyuk 7,000 years earlier, still had not died. The most important of them all was celebrated at Ephesus.

Here the River Kayster flowed into a gulf which formed a superb harbour. The settlement that grew around it thus linked the trade routes, coming from Persia and the lands of the Tigris and Euphrates beyond to the east, with the increasingly prosperous world of the Mediterranean to the west. In 1000 BC, Greeks themselves settled here and built a splendid city.

The city's connection with a fertility cult was implicit in its very name which, in its pre-Greek form, was more accurately represented as 'Apasus'. It was derived from a word meaning a bee, which has also given us, by way of Latin, our term 'apiarist' for a bee-keeper. The bee, which so industriously gathers nectar from wild flowers and turns it into rich stores of delicious honey, had long symbolised the bounty of nature, and Ephesus adopted the image of a bee as its emblem and used it on its coins. Bees and honey were also important elements in the offerings that were made to the mother goddess, the mistress of all wild nature, whose shrine had stood in the city since long before the Greeks arrived. The Greeks called her Artemis. The Romans, who took over the city from the Greeks at the beginning of

A conquest symbolised
The events of history are often told, many generations later, as stories in which single characters represent whole nations. This scene, painted on a Greek vase in the first half of the sixth century BC, vividly shows Theseus slaying the bull-headed monster, the Minotaur, but it may also be seen as a symbolic representation of Greeks from the mainland conquering Minoan Crete.

the Christian era, knew her as Diana. Her temple, the Artemision, became so rich and splendid that eventually it was listed as one of the seven wonders of the world. Her image appears on coins, but it is known to us in most detail from several marble copies of her statue, made during Roman times, that have been recovered from the city's ruins. In these, she is represented as she appeared at the climax of the festivals held in her honour, decked with offerings.

She wears a tall crown on her head. Her belt is decorated with the symbols of bees, and a long panel in the front of her skirt is ornamented with images of lions, bulls, goats and mythical winged creatures like griffons. Deer stand on either side of her. Both the panel and the crown in the original were probably made of gold and fastened on to her body which was fashioned from wood. The objects shown in the marble figures as hanging around her neck were real offerings. They include garlands of grapes and acorns, but the most prominent are three rows of large bulbous objects, shaped somewhat like eggs and varying in number from fifteen to thirty, that are slung around her just above the waist.

Since Christian times, these have been customarily described as breasts, and been regarded as symbolising, by their extraordinary number, the goddess's extreme fertility. During the Renaissance, when Europe's interest in antiquity was reawakened, Italian artists visualised her with milk spurting from all of these breasts. She was, in consequence, a highly suitable subject for a fountain, though hardly an image that induced reverence. However, the Renaissance artists did not know the Roman statues and when these are examined closely it is immediately clear that whatever else these objects are, they are not breasts. They resemble them neither in shape nor position, let alone number. Most important of all, they lack any signs of nipples. It is scarcely likely that the Roman sculptors who, elsewhere on the figure, carved clusters of acorns and images of bees in accurate detail, would have omitted such an essential feature. Later scholars recognised this and made several alternative suggestions as to their identity. They were gigantic grapes – in spite of the fact that accurately carved grapes hang in bunches above them; they were eggs – though how they could be strung in such a strange fashion was not explained; or maybe – in desperation – they were the fruit of egg-plants. None of these hypotheses accounted for the size and the shape of Artemis's strange garlands, nor the way in which they hang. Only within the last few years has their character been convincingly identified. They are the severed testicles of bulls, each pair still enclosed within the scrotum.

Castration, and the offering of testicles to fertility gods, has been recorded elsewhere – in Greece, in southern Italy and in Sicily. The importance of the bull god in the worship of Artemis at Ephesus has

The goddess of Ephesus
The importance of the fertility cult in Ephesus was emphasised by the city's use of the bee as its emblem on its coinage (*above*). Surviving images of the ancient goddess herself are of a comparatively late date. This one, which was unearthed in the city, was made by Roman craftsmen in the second century AD. It is of fine marble with traces of gold on its upper torso. The image that stood in the temple eight hundred years earlier was probably of painted wood. During festivals in the goddess's honour, it was dressed in splendid clothes, decorated with jewellery, and hung with the offerings and sacrifices that were made in her honour.

also long been recognised. Its occurrence here is hardly surprising, since Çatal Huyuk, with its innumerable bull shrines, lies less than 300 miles away to the east. Crouching bulls appear on the panel of the goddess's skirt. Bull heads are carved on columns in the city's market place. Bulls occur again and again in the decorations of monuments and tombs. When the ruler of the nearby kingdom of Lydia came to pay tribute to Artemis, he chose to present to her a statue of solid gold in the form of a bull. And now, recent excavations near her temple have put the connection beyond doubt.

The site of the last and grandest of the several temples that were erected in her honour has been known since the middle of the last century. It lies a little distance to the north-west of the ruined town where piles of marble masonry and sections of columns lie beside a shallow marshy pool. Thirty-five yards west of it, however, a team of Austrian archaeologists discovered the foundations of a hitherto unknown building, and buried within its floor, they found innumerable dismembered cattle bones.

The building was surrounded on three sides by high walls. The fourth side was formed by a narrow ramp, twenty yards long, that sloped down to a platform at one end. Beyond this stood a pedestal. The excavators believe that here bull sacrifices were made. They deduced that, during the annual festival of Artemis, the animals were tethered in a line on the ramp. One by one, they were led down to the platform and there ritually slaughtered and castrated. Their testicles were then fastened in rows on the wooden gold-clad image of the goddess that stood on the pedestal beyond the sacrificial altar. When all had been killed, the image, girt with her grisly trophies, was carried back to her place in the main temple where more garlands were hung upon her, while the remains of the bulls were buried in the yard of the altar building.

This cult was a famous and widely popular one. Pilgrims came long distances to pay tribute and brought considerable wealth to the city. Silversmiths, for whom Ephesus was famous, grew rich by selling images of the goddess to visitors. In AD 53 Saint Paul came here to preach the new religion of Christianity. But, as the Bible records, he was not allowed to do so for long. His message began to damage trade. Eventually the silversmiths fomented a riot in the huge theatre in the centre of the city, and the enraged mob went in search of Paul. Two of his companions were badly beaten, Paul himself was forced to flee, and Diana's cult continued for further centuries.

Signs of fertility at Ephesus
The temple in which Diana was worshipped is now totally ruined. One column has been re-erected (*right*) on the top of which, as is appropriate for the goddess of natural fertility, a pair of storks now regularly nest. In the Roman city itself, however, many fine buildings still stand, and in the details of their decorations, like those on the fountains in the baths (*above*), further evidence can be found of the importance of the bull cult.

THE BUTCHERY OF THE GODS

The Romans, when they succeeded the Greeks as masters of the Mediterranean, readily acknowledged local gods wherever they encountered them in their gigantic and expanding empire. By the first century AD this had spread across the whole of the Mediterranean world from Spain to Syria. So the practice of animal sacrifice, in the cause of maintaining the fertility of the natural world, not only continued but spread. Mithraism, another bull-cult which had its origins in Persia, became particularly popular among the Empire's soldiery. By the third century AD, it was at its height and rivalling Christianity in popularity. Its hero was the god Mithras. He was born, it was said, in a cave. In the forests outside it, he encountered a great wild bull which he took back to his cave and there slew. Its blood, gushing on to the ground, spawned all the creatures of the earth.

Mithraic temples were usually underground, reproducing the cave in which the creation battle took place. At one end stood an altar, and beyond it a sculpture or a wall painting showed the god plunging his dagger into the neck of the bull. In the sky above, a raven appears. Beneath, on the ground, the life-blood of the bull is lapped up by a dog and a snake, while a scorpion grasps the bull's testicles. The detailed symbolism of these unvarying protagonists is still not fully understood, for the cult demanded secrecy of its devotees and no comprehensive account of its practices survives. We know, however, that the worshippers in these underground chapels dressed themselves in animal costumes, and danced as lions and ravens before the altar. But one thing was certainly clear. Although the bull was still acknowledged to be the fount of fertility, a human figure has now been elevated to stand alongside it and force it to release its power. The change is a significant one.

Mithras, slayer of the bull
The worship of Mithras was usually conducted in dark subterranean temples like the one (*right*) discovered beneath the Christian church of San Clemente in Rome. The god himself is nearly always represented in the same costume and the same posture, surrounded by the same animals: the dog, the snake and the scorpion. The signs of the zodiac are also often displayed around him, as here (*left*) in a marble relief carved about 400 AD and found in Sidon.

Capturing animals
Hunting animals was a widely popular pastime in Roman times, but sometimes they were caught alive in order that the pleasure of witnessing their deaths could be shared by thousands in the arenas. This mosaic comes from a floor of a grandiose villa built during the fourth century AD near Piazza Armerina in Sicily.

The attitude towards animals implied by the Mithraic image was certainly widespread among the Roman public. Hunting was one of the most popular sports, and the straightforward slaughter of animals the most entertaining of spectacles. People scoured the countryside for bears, stags and boars and pursued them with spears and dogs. Those who could afford to do so went on long expeditions to the wilder parts of the empire in search of lions and hippopotamus, hyaenas, leopards and crocodiles, and lovingly recorded their exploits in the mosaics that decorated their luxurious villas, anticipating only too precisely the trophy rooms, hung around with the stuffed heads of slaughtered animals, erected 2,000 years later by the big-game hunters of the British Empire. Such extravagances, however, were beyond the means of the majority of people living in the densely packed tenements of the great Roman cities. To entertain them, animals were brought back alive to be tormented and killed in the public arenas.

The biggest of these that still survives is the Colosseum in Rome. There fifty thousand people could be seated on the stone benches that rose tier upon tier around the oval sand-strewn arena. But there were

Slaughtering animals
In the arenas, animals of all kinds, the rarer and the more ferocious the better, were slaughtered by professional gladiators, who became personally extremely popular with their own enthusiastic following. Their named portraits were often part of the decoration of the villas of the rich. This painting of one, Fulgentius, slaying a leopard comes from the Roman city of Leptis Magna in Libya.

many others that rivalled it. Thysdrus, in Tunisia, now known as El Djem, has one that is almost undamaged. It stands 120 feet high, towering like some huge cliff above the low buildings of the Arab town that cluster around its base. It could seat thirty-four thousand spectators.

The festivals held in these vast buildings were, in origin, religious. In earlier centuries they were mounted as thanksgivings to the gods. But as the Empire grew, their religious element diminished to vanishing point and the blood-letting grew into unbridled butchery. When the Emperor Titus inaugurated the Colosseum, there were 100 days of celebration and, on just one of them, 5,000 animals were slaughtered. To supply the arenas, units of the imperial army stationed in Europe were taken off their military duties to concentrate entirely on catching bears. Special animal-catching expeditions were sent to Africa. The demand was so great and the hunting so intensive, it may well have been a significant factor in the extermination from the Roman provinces of many larger animals such as lions and hippopotamus.

The caged animals were kept in underground dungeons that lay beneath the arena itself. On the day of the festival, they were hauled in their iron cages on to lifts that operated by counter-balances and rose up in the centre of the arena. The Colosseum had thirty-two of these ingenious devices. As the crowd roared with excitement, drums were beaten, trumpets blown and the terrified animals were released. Sometimes they were goaded to attack one another; sometimes men armed with spears and tridents pursued them around specially erected shrubberies, in a theatrical imitation of a real hunt in the open countryside. Lions and lepoards that had been specially starved for days were released into the ring and presented with a human slave or a prisoner of war lashed to a post, and encouraged to rip him to pieces, before they themselves were speared and stabbed by gladiators. The most titillating killing of all did not even involve animals. In these, men were forced to slaughter one another for the entertainment of the crowd.

Men become animals
The arena in El Djem in Tunisia (*right*) still has in its centre the underground corridor, now without a roof, that led to the cages where the animals were kept, together with the lift shafts on either side. Men fighting boars bare-handed and spearing deer, were common sights; but the most exciting spectacle was that of a human being – a slave, a prisoner of war or a criminal – being torn apart by an animal. That was such a popular event it was thought an excellent subject with which to decorate a villa. This example (*above*) in mosaic comes from Leptis Magna, in Libya.

This attitude to life was reflected in the Romans' view of the whole of the natural world. To them it seemed that nature could be ravished and plundered as men wished. Its products were self-renewing and inexhaustible. They saw no reason why men should not take what they wanted as often as they wanted it. The state gave legal title to undeveloped land to anyone who cleared it of forest. As the human population around the Mediterranean grew, so more and more of the forests that had once girdled it with green were destroyed.

The felling had begun long before, in Greek times. Plato, in the fourth century BC, had vividly described its progress on the hills around Athens. 'What now remains, compared with what existed', he wrote, 'is like the skeleton of a sick man, all the fat and soft earth wasted away and only the bare framework of the land being left.'

Wood was virtually the only fuel used in the classical world, and the vast proportion of the timber felled was burned. It was needed for cooking. It was used to heat the houses, whose flues and hypocausts are such an admired invention of the Roman architects. It had to fuel the kilns which produced bricks, pottery and tiles. It was an essential building material, for even though a house had stone walls, its roofs, rafters and doors were of wood. Chariots and carts were constructed from it. So, most importantly, were ships. When states went to war, entire forests were devastated to provide the armies with vehicles and the navies with ships. So, as the classical empires spread from east to west along the Mediterranean and north into Europe, the forests were demolished.

The consequences were most severely felt on the southern and eastern shores, where the rainfall was low. Here the forests had been a key factor in maintaining the health of the land. They absorbed the rain when it fell in winter, and retained it in the soil around their roots. In summer they released it slowly, so that the shaded land never dried out entirely, and springs flowed throughout the year. Their removal was catastrophic.

The provinces of North Africa were, originally, among the richest in all the Empire. Six hundred cities flourished along the African shore between Egypt and Morocco. The biggest of them, Leptis Magna, had a population of around a hundred thousand. The soil was so rich, according to Pliny, that one grain of wheat planted in it, would sprout and produce a stem bearing 150 grains. By the end of the first century AD, North Africa was producing half a million tons of grain every year and supplying the huge city of Rome, which had long outstripped its own agricultural resources, with two-thirds of its wheat.

The theatre, Leptis Magna
Here the entertainment was drama, music and dancing.

The end was not long in coming. There is still argument as to how much a change in climate contributed to the final collapse. The balance of opinion seems to be that, though rainfall did diminish, the crucial blow was the stripping away of the trees and the relentless ploughing and reploughing to extract the maximum tonnage of crops.

Year after year the soil of the fields was lost. In summer it was baked by the sun and blown away by the hot winds. In the winter, rain storms swilled it away and rivers carried it down to the coast and deposited it in their deltas. By the fourth century BC, the port of Ephesus had silted up so badly that it had to be abandoned and rebuilt on a new site farther along the coast. Here, once again, the harbour began to fill. Regulations prohibiting the sawing of wood on the jetty, so that sawdust would not clog the channels, were of no consequence. Even dredging could not keep the harbour clear. Eventually, ships of any size were unable to reach the quays and the city lost much of its trade. Today, the harbour is separated from the sea by three miles of flat, marshy land.

Thermopylae, on the Greek coast, was the site in 480 BC of one of the most heroic battles in ancient history. A tiny detachment of Greek soldiers, commanded by the king of Sparta, held a narrow pass between the sea for three days against a huge Persian army. Today, that pass no longer exists. The soil from the hills above has been washed down by the rivers and deposited at the edge of the sea in such quantities that the pass has been transformed into a wide plain.

All along the African coast, the land dried out. Wheat could no longer be grown; olives, which had once been prohibited by law lest they should displace the more highly valued wheat, were the only crop that would grow. Then even they began to fail. The human population dwindled. Sand blew through the stony fields and the grandiose buildings tumbled into ruins. Today, the harbour at Leptis, where once great ships came to fill their holds with grain, is buried beneath sand dunes.

It used to be said that such cities collapsed because here nature failed to support man. The truth is the reverse. Here, man failed to support nature. The bull, the all-powerful god of fertility who had been worshipped since long before men began to build cities, was now dethroned. The last debased relics of his cult were to persist and resurface later in the ritualised slaughter of the Spanish bull-ring. But for the most part, he lingered among the ruins of the classical empires not as a god, but castrated, enslaved, and yoked to the plough.

Roman Ephesus – a stranded city
The avenue from the theatre once led straight to the port.
Now the harbour is silted up and the sea three miles away.

PART THREE

THE WASTES OF WAR

Oxen pulling a plough in Portugal.

THE TAMING OF THE HORSE

The Roman Empire was gravely weakened when deserts overwhelmed its African territories, but the invasion that brought about its final collapse came from the other side of the Sea, to the north. And just as one animal, the domesticated bull, had played a key part in the creation of the wealth of the empire, so now another, the horse, became instrumental in bringing about its destruction.

The Romans had long known of the nomadic tribes who roamed the open plains beyond their frontiers in Hungary, Poland and the Ukraine. They called them, contemptuously, *barbari* – barbarians, meaning people who spoke neither Greek nor Latin but could only manage an incomprehensible stuttering gabble – *ba-ba-ba*. These tribes were perpetually on the move, the men on horseback, driving their livestock before them, the women and children following behind in waggons. They never slept inside a house; they had no knowledge of the plough; they lived entirely on milk and meat. Their cruelty shocked even the Romans. After a fight, they flayed their victims and slung the bloody skins over the backs of their horses as trophies.

Their dependence upon their horses was total. A Roman historian, Ammianus Marcellinus, wrote 'They are no good at all at fighting on foot, but perfectly at home on their tough and ugly horses. They even ride side-saddle when they relieve themselves. It is on horseback that each of these people buys and sells, eats and drinks and, bent across the neck of his steed, takes a deep sleep filled with pleasant dreams.' A Roman poet put the matter more succinctly. 'The cloud-born Centaurs are not more closely joined to the bodies of horses than they.'

The barbarians had, for centuries, been raiding the Roman settlements along the northern frontier, but towards the end of the fourth century AD, their behaviour began to change. A thousand miles away, around the northern shores of the Caspian Sea in the far-off steppes of Russia, a particularly war-like group called the Huns started to move steadily westwards. As they travelled into Europe, the tribes close to the Roman frontier moved south and west to avoid them. Now, when barbarians raided a Roman settlement, they did not return east to their

Noble Romans defeat uncivilised barbarians
This marble relief forms the side of a sarcophagus and represents a battle between the barbarians and the Roman army, an increasingly common event during the 3rd century AD when it was carved. The sympathies of the sculptor are plain. The Roman soldiers surrounding their mounted commander, who may be their Emperor, are shown as men of great bravery and serene nobility who, without exception, are vanquishing the half-naked, shaggily-bearded barbarians whose faces are contorted with anguish. The true outcome of such battles was often very different.

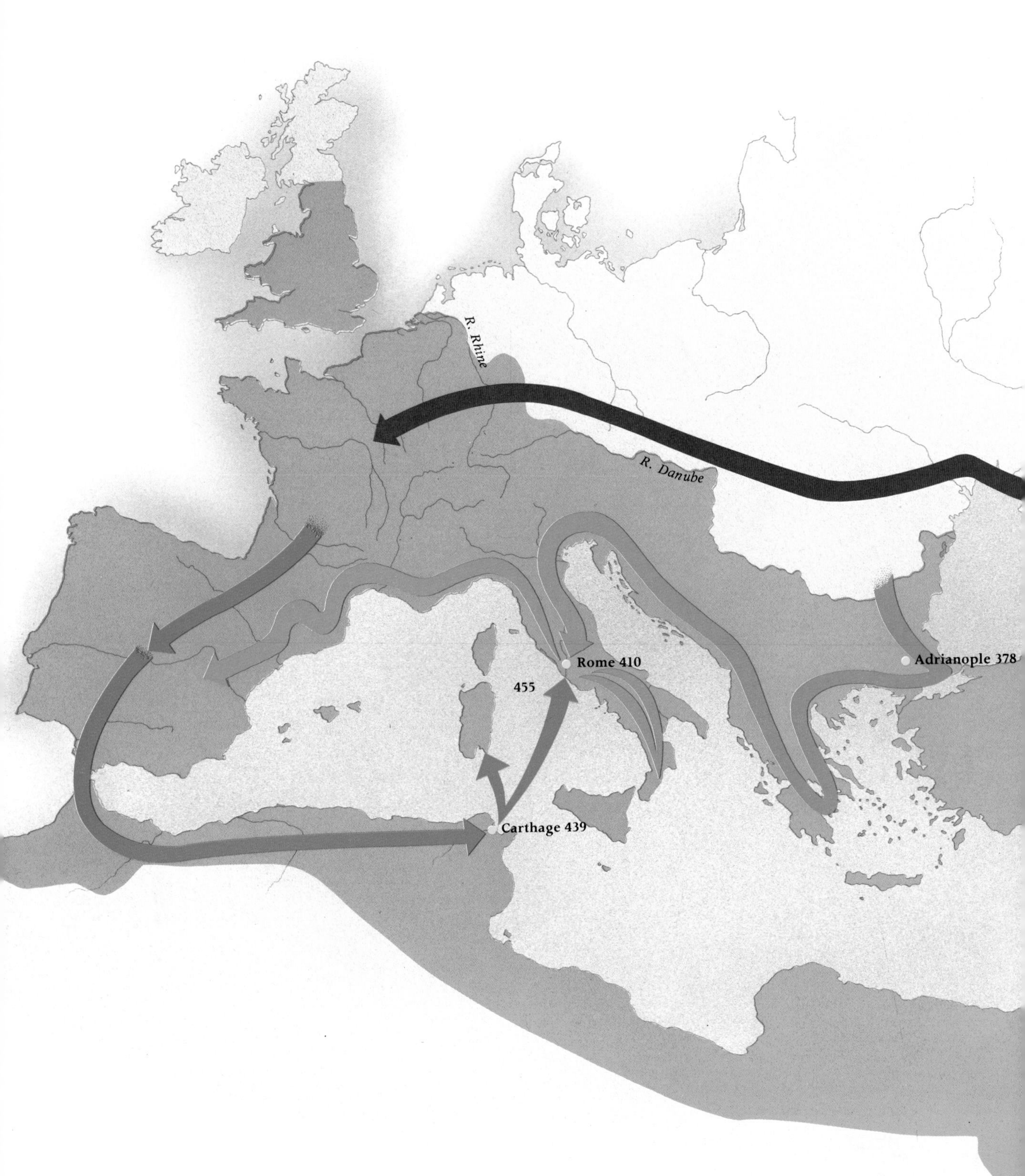
R. Rhine
R. Danube
Rome 410
455
Adrianople 378
Carthage 439

own territory but rode on deeper into the lands of the Roman empire. As time passed, these undisciplined bands coalesced into an army, and the random search for booty turned into a mass migration that was to continue for generations.

The Visigoths, who had been living on the shores of the Black Sea around the mouth of the Danube, moved south into Roman Bulgaria. In the year 378, they came face to face with the Roman army at Adrianople, today's Edirne in Turkey. The Visigoths won decisively, killing the Roman emperor and two thirds of his men. With little now to stop them, their chieftain, Alaric, led them into Greece and on round the shores of the Adriatic to lay siege to Rome itself. They took the city in 410, sacked it and moved on around the Mediterranean coast of France until they reached Spain where they at last settled.

The Huns themselves continued to advance westwards, keeping well north of the Mediterranean, and rode across Germany into central France. Other tribes, having heard stories of the lush living that could be had on the sunny shores of the Mediterranean, were drawn southwards. The Vandals fought their way down from north-western Europe through France and Spain. Lured by the long-standing but now outdated reputation of north Africa as the granary of Rome, they crossed the Gibraltar Straits and swept through Morocco, Algeria and Tunisia to occupy the city of Carthage in 439.

The Roman army had little defence against these ferocious horsemen. Its strength was its superbly disciplined infantry, marshalled into legions. The few horses it possessed were big, heavily built animals which were largely reserved for officers. They were well suited to carrying a fully-armed commander, but they were clumsy and difficult to manoeuvre when attacked by the barbarians on their small, agile horses, who came racing over the horizon at full gallop.

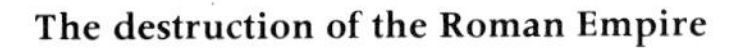

The wild horse of central Asia
These Przewalski's wild horses are part of a herd living in a zoo. It is very doubtful if any individuals still survive in their original home, the steppes of central Asia, for none has been seen there since 1968.

Wild horses had once lived over most of Europe. The last herds of them were discovered in 1879 by the Russian naturalist Nikolai Przewalski in the remote semi-desert of Mongolia, and the species was accordingly named after him. What he saw, known locally as the taki, was a brownish animal standing about $4\frac{1}{2}$ feet at the shoulder. It had a massive, heavy-jawed head, a thick neck and a short, erect, bristly mane. Two other subspecies were later identified: the tarpan, which lived on the grasslands of the Ukraine and was greyer in colour, with a prominent black stripe running along its back; and another similar but slightly smaller, lighter-coloured animal that lived in the forests.

Man had hunted horses for food ever since he had first encountered them, but he did not manage to domesticate them until some three thousand years after he had tamed cattle. Identifying the time and place of that important achievement is not easy, for horses changed very little anatomically during the early stages of domestication. The clearest evidence comes not from their bones but from objects made by man to control them. The earliest of these so far discovered were excavated from a grave in the southern Ukraine, dated around 3500 BC. They are six small pieces of antler bone that seem to have served as the cheek pieces for bits.

Initially, the domesticated horses kept in such settlements were undoubtedly eaten, as their wild ancestors had been. Leg bones broken open for their marrow and skulls split for the extraction of the brain are evidence of that. But there are also clues showing that soon they were exploited in other ways.

If horses are kept throughout the long cold months of winter, when the grass on which they normally feed has died away, they have to be given special fodder. This is laborious to gather, and it has to be stored in considerable quantity if it is to last throughout the winter. Animals that are kept only for their meat are therefore likely to be killed in the autumn, and so the bones in the kitchen refuse will be predominantly those of younger animals. If a high proportion of bones from older animals is found around a living site, it is a strong indication that the horses there were kept for other reasons than for their meat. And this is what was discovered in the Ukraine sites. Whether at first this other purpose was for riding or for pulling waggons is not certain. Cattle were being worked as draught animals at this time, so the possibility of using horses in the same way must have been an obvious one. Even so, it is difficult to imagine how herds of horses could have been managed and controlled unless men were able to ride at least some of them. But the first person who sat on the back of a rearing, bucking, kicking, biting wild horse must have been brave indeed.

Once that great step had been taken, however, and the horse had become humanity's servant, the lives of the people were transformed. Their carts and waggons could now move twice as fast as the lumbering oxen had managed to pull them. More importantly, human beings were no longer outpaced by every other large animal on the steppes. Now they could gallop to the farthest horizon to chase a deer or to repel an enemy. Now they could travel sixty miles in a day. No wonder the barbarians seldom dismounted.

The techniques of selective breeding had been worked out during the preceding millennia as the people steadily moulded the size and proportions of cattle and sheep to suit their purposes. Now the same methods were used for breeding horses. Big, strong animals were developed for pulling waggons; smaller, more agile ones for riding. Early in this process a genetic change occurred, perhaps by accident. The domesticated horses no longer shed the hair of their manes every year, as they had done in the wild, but retained it, so their bristly upright manes became long and flowing.

Over the next thousand years the practice of taming and breeding horses spread west from the steppes to the shores of the Mediterranean. The use of the horse in war, however, required consummate skill. Saddles and stirrups had not yet been devised, and wielding a sword or hurling a spear when riding bare-back was far from easy. By

1800 BC, tribes in Syria had harnessed their horses to small chariots with spoked wheels which, because they were so light, could be driven at speed. This gave the Syrians a swift-moving platform, and a warrior could brace himself against its walls in order to handle his weapons. Equipped with this new vehicle, the people from the eastern end of the Mediterranean went to war against Egypt. The Egyptians called these invaders the Hyksos, a name that meant no more than 'rulers from foreign lands'. Several times the Egyptian state was conquered by them, but each time the native-born Egyptians eventually regained their independence. And from the Hyksos the Egyptians acquired the skills of charioteering. Their vehicles, made largely of wood and leather, were drawn by two horses, one on either side of a long shaft, and carried two men, a driver and a warrior armed usually with a bow and arrows.

In due course, the Greeks copied these techniques from the Egyptians, but their rocky hilly land was not ideal country for chariots; initially at least, they used them not so much for war as for sport. Chariots with four horses harnessed abreast were raced during the Olympic Games in 680 BC. Horses with riders did not compete against one another until over thirty years later in 648 BC, and horsemen did not become an important element in the Greek armies until three hundred years after that.

The first disciplined and manoeuvrable troop of cavalry was brought together by King Philip II who ruled over Macedonia, a small state to the north of the Greek peninsula. He fought his neighbours for control of the whole of Greece and in 338 BC met a much superior army from the south at Chaeronia. With one carefully timed charge

Harnessing the horse
When these bronze panels (*far right*) were made for the doors of the palace of an Assyrian king in Iraq, during the 9th century BC, the techniques of yoking horses to chariots had been known and practised for a thousand years. In contrast, the equipment necessary for riding a horse was still little developed. The Greek warrior of the 6th century BC, represented by the bronze statuette (*right*), has neither saddle nor stirrups.

of his cavalry, two thousand strong, he won the day. The attack was led by his eighteen-year-old son, Alexander, riding a splendid black stallion with a white star on its forehead, named Bucephalus – a complimentary name which means, interestingly, 'Bull-headed'. After Alexander succeeded to the throne, Bucephalus carried him far into Asia on his twelve-year campaign. At the last great battle in India, on the Jhelum River, the stallion was wounded and died shortly afterwards at the age of twenty-three. Alexander, heart-broken, buried him and built a town around the tomb in his horse's honour, naming it Alexandria Bucephali.

Skilled horseman though Alexander was, he, like the riders who had been carved on the marble frieze of the Parthenon a century earlier, still rode without either saddle or stirrups. Those devices were to come, as domestication itself had come, from the barbarians of Central Asia. The earliest evidence of the use of stirrups may be a painting on a vase recovered from the grave of a chieftain who was buried near the Dnieper River in the Ukraine around 300 BC. It appears to show a horse with a saddle on its back, secured around its belly with a girth, and with a leather loop hanging from it in which the rider might have put his foot. If this is indeed a correct interpretation, then it took a very long time before the stirrup was generally adopted, for the barbarian horsemen from these plains were still riding without stirrups or saddles when they galloped out of their homelands in the fourth century AD to demolish the Roman Empire.

Riding the horse
The nomads of central Asia, represented on this silver gilt amphora (*right*) made between the 6th and 4th centuries BC, used saddles secured to the horse's back by a girth as well as a bridle, bit and reins, and prevented their horses from wandering by hobbling them. But neither they, nor Alexander the Great in the 4th century BC, shown above in a Roman bronze statuette, had stirrups to give them a firm seat from which to fight.

THE TRIUMPHS OF ISLAM

Animals that live in hot deserts tend to be smaller, slimmer and with relatively longer legs than their cousins living in colder climates. The reason is connected with the rate at which different shapes and sizes lose heat. Large bodies and stout legs tend to retain heat and this suits mammals living in the chilly north. The forces of evolution tend, therefore, to produce these proportions in animals living there. Small elongated shapes, on the other hand, lose heat quickly, and that is of great benefit to animals in warm countries where overheating can be a real hazard. Such differences between populations in cold and warm climates can be seen in mammals of all kinds – rabbits and camels, mice and men. It is also found in horses. Horse breeders also try to produce animals that suit the local conditions and so reinforce the effect of evolution by natural selection. Accordingly, the breeds that developed in the Levant and North Africa were small, agile and long-legged compared with those more massive horses of northern Europe. Slimness and long legs lead to nimbleness and speed, and the horses of Arabia became among the swiftest of all. It was these high-spirited animals that carried the warriors who built the next great empire around the Mediterranean.

The most favoured of horses
The Arabs' devotion to their horses is legendary. Sheiks can recite from memory pedigrees of their horses that stretch back to the times of Muhammad. Certainly, the horses that came from the Arabian deserts are among the swiftest, liveliest and most beautiful of all. For centuries they have been sought after by breeders, and their genes are now carried by horses in studs all over the world.

In the sixth century AD the semi-nomadic tribes living in the deserts of Arabia had no common language nor religion and were largely independent of one another. Their settlements stood on the caravan routes that ran along the Red Sea coast, linking the Yemen in southern Arabia with Syria and the rest of the Mediterranean world. Baggage trains of camels and horses trudged along these ancient trackways carrying loads of gold and spices and, in particular, frankincense, the resin of a kind of balsam tree that grew in Somalia, which was greatly valued for the perfume it created when burnt. The richest of these towns, standing mid-way along the coast, was Mecca. Here in the first quarter of the seventh century a new religious leader, Muhammad, began to preach. As a young man, he had travelled with the caravans, working as a camel driver. He had married a wealthy widow and become rich himself and, when he was forty, he began to see visions and receive divine revelations. He proclaimed that there was only one god, Allah, and that the multitudes of spirits and idols worshipped by other Arabian tribes were false. He urged his listeners to submit themselves to the will of Allah. Both the name of this new faith, Islam, and that of someone who follows it, Muslim, come from an Arabic word that means 'to submit'.

Muhammad's declaration of the rules by which Muslims should order their lives were collected, together with other texts from later religious writers, in the holy book of Islam, the Koran. It provides guidance for all aspects of human existence, and is particularly rich in information as to how horses should be cared for and how they

should be bred. 'For every barley corn given to a horse', it says, 'Allah will forgive a sin.' Muhammad himself was certainly a competent horseman, as might be expected from someone who had worked with the caravans, and he is credited with founding the breed of Arab horse that is known today. It is said that he deliberately kept a herd of horses in a compound without water for a week. On the seventh day of their confinement, he opened the gate to allow them access to a water trough. As the thirsty animals raced out, a martial trumpet sounded the call to battle. Five of the mares in the herd immediately turned back in response without drinking, and these became the foundation of his stud. All pedigree horses are reputed to trace their descent back to them.

The story illuminates not only Muhammad's love of horses, but his martial character and that of Islam. He and those who succeeded him as leaders of the faith preached that it was the duty of all believers to spread the word of Allah. Infidels who persisted in their disbelief should be punished in this world as they would assuredly be in the next. Those true believers who were killed while carrying the word of God in a holy war, a *jihad*, would be taken directly to Paradise. The faith spread like wildfire through the tribes and for the first time unified them. Under Muhammad's leadership and inspired by religious fervour, they set about claiming neighbouring lands for the faith. Other Arabic people – pagans in the eyes of Islam – were required to adopt the new religion on pain of death. Christians and Jews, who like Muslims believed in a single all-powerful God, were allowed for the most part to continue to practise their own faith, though the lands in which they lived had to submit to the rule of Islamic law.

The Levant at this time was still part of the rump of the collapsed Roman empire, ruled by the Byzantine emperor from Constantinople. Its conquest was one of Muhammad's earliest objectives. He was assembling an army to attack it when suddenly in 632 AD he died. Eventually, however, when his successor had established his command, the campaign was launched. The warriors who fought in the Islamic armies were well accustomed to the nomadic life and were expert at living off the country. They had no need of baggage trains to carry their stores. Nor were they accompanied by foot soldiers who would inevitably have slowed them down. Over the next few years they drove the Byzantine soldiers out of the Levant. Other armies were despatched westwards along the North African coast. The Berbers, pagans who lived in the Atlas Mountains, were subdued and converted to the new faith. Most of the Christian population, descendants of the Roman settlers, left Africa and took refuge in Sicily and Italy. By the end of the century, Islam controlled all the eastern and southern coasts of the Mediterranean from Syria to Morocco.

The cavalry of Islam
As bands of pilgrims approached the Holy City of Mecca each year, four days were spent in feasting and exchanging gifts; and horsemen, led by mounted musicians playing trumpets and beating kettle drums, paraded with banners proclaiming 'There is no God but Allah'.

In AD 711 Islam moved into Europe. An army of Berbers, led by Tariq, an ex-slave who had risen to be Governor of Tangier, crossed the Strait that the classical world had known as the Pillars of Hercules. They called the spectacular rock peak on the northern side Tariq's Mountain – *Djebel-el-Tariq* – which name became softened and shortened into Gibraltar. Their progress through Spain was swift even by Islamic standards. The local people in Spain had little affection for their aristocratic rulers, the alien Visigoths, and by AD 712 the country was largely in Arab hands.

The builders of this great Islamic empire, which at its height stretched from the Pyrenees to the Pamirs, came for the most part from simple desert-living communities. But they were followed by settlers who brought with them skills and scholarship from all over the Arab world, and under the new Islamic rule, ancient European cities blossomed into societies of dazzling sophistication. The Moors established their Spanish capital at Cordoba in the south of the country. They demolished the Christian basilica and using columns and masonry from the Roman ruins that littered the city, they built a mosque. It remains one of the most spectacular of Islamic buildings, a forest of over a thousand pillars connected with arcading and supporting a vast roof beneath which the faithful prayed and meditated and debated the affairs of the community. They installed street lighting. They constructed a sanitation system for the city and built

three hundred public baths – an innovation in Europe where hitherto little importance had been attached to personal cleanliness. They founded a university where, as well as theology, the Arabic sciences of astronomy, medicine and chemistry were taught. Their mathematicians introduced the concept of zero, which they themselves had acquired from Indian scholars, and worked with Arab numerals, displacing the clumsy unwieldy ones used by the Romans. Arabic words like algebra, azimuth, and amalgam were used, with all the bodies of knowledge that they imply.

The Arabs also brought a new attitude to the natural world. Coming from a desert country, they were skilled in techniques of irrigation, and they cherished plants with a fervour that was quite new to Europeans. Their gardens, encircled by high protective walls to keep out the searing winds, were blessed places indeed. Outside them lay hostile, barren sands and relentless, scorching sun. Inside, beneath the branches of fruit trees, there was cool shade, the perfume and colour of flowers, the songs of birds and the gentle plash of fountains. So a garden became identified in poetry, as in religious writing, with Paradise. Indeed, the same word in Arabic is used for both. So for them the very act of gardening was invested with a religious significance.

Islam in Spain
The Moors brought to Spain their own characteristic architectural styles, exemplified by the Great Mosque in Cordoba (*above*), and used their skill in controlling water to create gardens of a kind quite new to Europe, such as those of the Generalife (*right*) outside the walls of the al-Hambra, their fortified palace in Granada.

The Moors now began to create gardens such as these in Spain. The main axis of the garden was a long straight water channel. Paths ran beside it, and sometimes secondary streams flowed into it. In places the water was made to dance and play in cascades and fountains, catching the sunlight, moistening the air, filling the garden with delicate sound. Here the Arabs grew plants that were new to Europe, which they themselves had imported from countries farther to the east. They planted trees from China that produced delectable fruit – oranges, lemons and peaches. They cultivated strange vegetables too – carrots from Afghanistan which had not at this time acquired their yellowish-red colour and were, instead, purple, spinach from Persia, and aubergines, the name of which, in spite of its French sound, comes from the Arabic *al-badinjam*, which itself is derived from a Sanskrit word meaning, according to some authorities, an anti-flatulent vegetable.

The most famous of their surviving gardens are those they built around the al-Hambra, the Red Palace, the fortress that overlooks the city of Granada. Although much has been changed in the five hundred years since they were created, their basic plan, a succession of shaded perfumed courts where water slides and gurgles along wide shallow channels, remains sufficiently the same for a visitor even today to sense something of the enchantment that such places must have had for the people of Europe who still had not emerged from the cultural darkness that followed the collapse of Roman power.

The Arabs also brought with them new skills in handling birds. Peacocks, the most dazzling and spectacular birds in all the world, were imported from Islamic territories in India to strut and shriek in the gardens of the Caliphs. They also kept pigeons with great success, using them both for sending messages and for food.

The pigeon had great value to Europeans. The practice of killing livestock in the autumn, because of the difficulty of providing winter fodder, was still general. Techniques of preserving meat were unknown. The best that could be done was to dry it and salt it, but after a few months it became so rotten and rank in taste that it had to be heavily disguised with spices. Such problems, however, did not arise with pigeons. For most of the year, they fed themselves, flying over the countryside to glean tiny particles of food from the fields. They could even survive the bleakest weeks of winter with no more than a few handfuls of grain. So they constituted a source of fresh meat that could be taken at any time of the year. Furthermore, they laid as many as a dozen clutches of eggs every year, which also made excellent food, and their accumulated droppings formed a rich fertiliser for the fields.

Pigeons were probably the first of all birds to be domesticated. The Egyptians and the Greeks had kept them. The Romans built special towers, *columbaria*, in which the birds were imprisoned and fattened on a diet of chewed bread. They were prevented from losing weight through exercise by having their flight feathers pulled out and even their legs broken.

The Arabs, however, kept them in a different way. Pigeons in the wild nest on cliff ledges and the Arabs provided them with man-made substitutes in the form of great towers. The walls were made from stacks of short tapering pipes plastered together so that their mouths faced either inside or outside the tower. The birds readily nested inside the pipes and made the towers their permanent homes throughout the year. Such buildings in their most splendid form are grouped together to create a multi-turretted miniature castle surrounded by a perpetual halo of fluttering birds. They can still be seen today in Egypt, even though they have fallen out of fashion in Spain.

Falconry too was a favourite pastime among Arab people, as it still is. Europeans knew of this sport, but lagged well behind the Arabs in its technique. One problem faced by a falconer is the need to prevent a bird from taking off and chasing prey before the falconer wishes it to do so. This was dealt with by piercing the bird's eyelids which, as in all birds, rise up from the bottom of the eye, and inserting a thread

The skills of handling birds
Domesticated pigeons are undemanding birds. In Egypt many villagers keep them for food and provide them with castle-like dwellings of mud-brick (*below*) in which the birds nest and breed almost continuously. Falcons are altogether more highly-strung and temperamental; but the Arabs, early in their history, mastered the skills necessary to use them for hunting, and the sport still remains extremely popular today. These birds (*right*) belong to the Sheikh of Dubai.

through them so that when the threads from both eyes are tied over the bird's head, the eyelids are drawn up and the bird is effectively blinded. Arab falconers, though still using this technique of 'seeling' in some circumstances, provided their hunting birds with hoods that had the same effect. This was more difficult than it may sound, for no light whatever must penetrate the hood if the bird is to remain calm. The hood, therefore, has to be a perfect fit and be closed tightly around the beak with two drawstrings.

As well as these skills in handling animals and plants, the Arabs had a penetrating curiosity about the natural world around them. In part this was an inheritance from the writers and philosophers of classical antiquity, whose works in manuscript were still treasured in the libraries in the eastern part of the Islamic empire. But to this they added their own experience and genius. It was the Arabs who experimented with and codified the use of plants for drugs; they who wrote the earliest medieval manual of veterinary science; and they who between the eighth and tenth centuries raised scientific investigation of the natural world to the highest level it had yet reached around the Mediterranean.

Elsewhere in Europe, many of the attitudes to nature were still based on beliefs and superstitions that dated back to ancient pre-Christian times. The mandrake plant, for example, was credited with supernatural powers of the most dramatic kind. There was good reason to regard it with considerable awe, for its root, leaves and fruit all contain a powerful narcotic drug which can dull the senses and even induce hallucinations. Its root, which is like that of a small parsnip, is often cleft in two, so that the whole plant can be seen as a small human figure with its root as a pair of legs and the leaves a shock of hair. Close inspection might even reveal whether an individual plant was a mandrake or a womandrake. Indeed, it was often drawn as a small homunculus in manuscripts about medicine and witchcraft.

Pulling one up was a dangerous business. As the plant was dragged from the ground, it screamed and anyone who heard that dreadful cry would be struck dead. There was, however, a way in which it could be gathered without risk to human life. The act should be done on a dark, moonless night. The herbalist performing the operation had to plug his ears with beeswax and take with him a trumpet and a dog. After cautiously loosening the root with his sword, he had to tie the dog's lead around the root. Then, on the stroke of midnight with his back to the wind and blowing the trumpet to keep the mandrake's lethal shriek from his ears, he should whip the dog so that it bolted. The root would be pulled out, the dog would die, but the herbalist could then take the mandrake away in safety.

The salamander, an innocuous amphibian like a big newt, was also regarded with a mixture of horror and awe. It is certainly one of the most dramatically coloured animals in the European countryside, being black blotched with a brilliant golden yellow. As it is an amphibian, its skin must remain moist if it is not to die, so it spends most of its time concealed beneath stones or under leaves and moss, and normally emerges at night. Only after a heavy storm is it likely to appear during daylight and be seen by casual observers. For this reason, it seems to have become associated with wetness and cold, and thus came to be credited with the ability to quench fire. This reputation certainly goes back to ancient times. Pliny, the Roman naturalist in the first century AD, heard of it and in the down-to-earth,

Pulling up the mandrake root
The artists who illustrated medieval herbals usually drew plants with considerable accuracy, but the myths surrounding the mandrake so clouded their vision that they regularly gave it human features. This illustration comes from a German herbal written around AD 1500, and it shows the approved method of uprooting this supposedly lethal plant without being killed.

59
Mandragora
Alraun

practical manner typical of the Romans, tested the proposition experimentally. He took a salamander and put the unfortunate creature into a fire. It was, of course, burnt to a cinder, and Pliny duly recorded the fact in his great natural history.

But that kind of scientific evidence was forgotten in northern Europe when the Romans departed and the salamander was still widely believed to be able to put out fires. Even today the black and yellow salamander is called the 'fire salamander'. It was credited with other spectacular powers as well. It does produce from glands in its skin a bitter-tasting secretion and its brilliant colours serve to advertise its unpalatability to other animals that might interfere with it. But in the Middle Ages it was believed to possess a venom of stupendous power. A thirteenth-century manuscript stated as sober historical fact that four thousand men and four thousand horses of the army of Alexander the Great were all killed because they drank from a stream through which a salamander had recently passed.

Such stories and descriptions about animals were collected and preserved in manuscripts known as bestiaries. They had no particular author but were compendia of scraps of information gathered from dimly remembered classical writers, from local folk tales, but least of all, it would seem, from first-hand observation.

It is hardly surprising that some of the descriptions of animals from far-off continents, which the writers cannot themselves have seen, are fanciful. The griffon was thought to live in the northernmost parts of the earth and to have a body like a lion and the wings and beak of an eagle. Arabia was the home of the phoenix, a bird that lived for five hundred years. As it approached the end of its allotted life-span, it built itself a funeral pyre, scented it with branches of frankincense and then turned to face the sun. As it did so, it burst into flames. When the fire had burnt out, a new phoenix arose from the ashes fully rejuvenated for its next five hundred years of life.

Some of the fantastic creatures described by the bestiarists appear to be attempts to explain animal curios that were traded from distant parts and the subject of great wonderment. The eight-foot-long tusks of the narwhal, straight, elegantly tapered and marked with spiral grooves, were thought to be the horns of the unicorn, which had the body of a horse and which could be caught by maidens in whose lap it willingly laid its head. The horns, therefore, were much treasured for their magical and aphrodisiac properties.

The fire salamander
This amphibian, so feared during the Middle Ages, occurs over most of southern Europe. Its patterns vary greatly, and some individuals are almost entirely yellow.

But even common creatures of the countryside were credited, like the salamander, with extraordinary powers as a result of misunderstood observations. Bears give birth to their young at an extremely early stage of their development, and a newborn cub is nearly naked, blind and no bigger than a rat. The bestiarists maintained that the mother bear actually licked her young into shape. That perhaps is an understandable misinterpretation. On the other hand, the beaver was said to emasculate itself. Its testicles were believed to be of great medicinal value. When one was chased, it would turn round, castrate itself with long chisel-like teeth and, while its hunters gathered the booty, make its escape. Not only that, but if another hunter then pursued it, the animal, as it fled, would lift up its rear to demonstrate that it had lost its masculinity and that there was no point in hunting it further – a story that any beaver hunter would have known to be false, if only because beavers do not have external testicles but carry them within their abdomen.

The ibis was believed to use its long curved beak to clean out its own bowels. The fox was said to catch birds by rolling in red mud and lying motionless so that birds would mistake it for a bloody carcass and come down to feed on it. Mice were thought to be generated from damp soil. Bees were born within the corpses of cows, wasps in donkeys, and hornets in horses. There was some confusion about weasels. Some authorities maintained that they conceived through their ears and gave birth through their mouths; others were sure it was the other way about.

The scribes who copied these manuscripts were usually monks working within a religious institution, and they lost no opportunity of drawing a lesson or a moral from what they wrote about. Indeed, the chance to do so seems to have been one of their main motives in studying natural history. They did not wish simply to describe, but also to discover the reasons why the Almighty should have created such animals and put them on earth. So the pelican, which they maintained fed its young with its own blood by pecking a wound in its breast, was regarded as a lesson in piety. The eagle provided a parable of the resurrection, for when it grew old, it flew high into the sky beyond the sight of men, and approached the sun so closely that its worn feathers scorched clean and the fog in its aged eyes evaporated; whereupon it dived down to earth into a spring of fresh water to emerge, like the phoenix, rejuvenated.

Perhaps we should not be too condescending about the uncritical attitudes of the bestiarists. One of the least fanciful of their descriptions was that snakes had damp bodies and left slimy trails. The slightest encounter with a snake is enough to prove to anyone that this is not so. Yet plenty of people, even today, still believe it.

MYTHICAL CREATURES OF THE MEDIEVAL WORLD

The illustrators of medieval bestiaries obligingly turned myths into images and drew the two-headed amphisbaena (*1*), geese emerging from goose barnacles (*3*), and the eagle (*2*) which, having scorched its wings by flying too close to the sun, plunged featherless into water and emerged rejuvenated.

1

2

3

The griffon (*1*) was part lion, king of the beasts, and part eagle, king of the air, and so it was king of all. Hedgehogs (*2*) collected apples on their spines. The phoenix (*3*), as it approached death, built its own pyre from the flames of which it was reborn; and the salamander (*4*) passed unscathed through fire. Opposite, the fox (*5*) caught birds by feigning death, the bear (*6*) had to lick its young to give them shape, and the cockatrice (*7*), whose very glance was lethal, could be killed only by the weasel which, though victorious, would inevitably also die in the battle.

1

2

3

4

5

6

7

THE MIGRATIONS OF THE CRUSADERS

At the end of the eleventh century, the rising tide of humanity began to flow once again round the Mediterranean. This time it moved in the opposite direction, from west to east. In November 1095, Pope Urban II held a great ecclesiastical conference at Clermont in southern France. At its conclusion, he preached a sermon to a huge congregation gathered in an open space in front of the city gates. The presence of the Muslim armies in the Holy Land, he declared, was an affront to Christianity. All true Christians should take up arms and march to Jerusalem to free it from the defiling grasp of Islam. His message was remarkably similar to that preached by Muhammad five hundred years earlier. Like the Islamic *jihad*, the Christian Crusade was to be a Holy War; and just as a slain Islamic warrior would be carried immediately to Paradise, so a fallen Crusader would be forgiven all his sins and go directly to Heaven. The Pope's message spread like wildfire, as Muhammad's had done. All over Europe the battle cry echoed through chapels, monasteries and cathedrals.

The peasantry responded only too quickly. Famines had ravaged Europe during the previous few years and many people were already leaving the land to look for a living elsewhere. If they were going to march to Jerusalem, they might as well start straightaway rather than labour in the fields to plant crops that they would never reap. So early in 1096, within a few months of the Pope's proclamation, a raggle-taggle army assembled, led by a number of itinerant barefoot preachers. As they tramped across Europe, so their numbers grew. Peasant families loaded their possessions on to ox-carts and joined the throng. Beggars, thieves and young men looking for adventure marched alongside the devout. They carried no supplies, and made violent demands for food and money from all the towns through which they passed. In central Germany, their religious fanaticism and mounting thirst for excitement and violence were vented on Jews. Great numbers of these people had been dispersed throughout Europe, driven out from their homeland in the Levant by the repeated subjugation of their country by foreign powers, starting with the Assyrians in the eighth century BC, and culminating with the Romans in the first century AD. They were often reviled as the murderers of Christ, and here in the cities of the Rhineland the would-be Christian pilgrims rampaged through the Jewish quarters, looting and killing.

More murderous riots broke out again as they passed through Bamberg and Prague. By the time they reached Constantinople, their plundering and extortion had become a habit, and there too they terrorised the citizens and ransacked public buildings. The Byzantine Emperor, desperate to get rid of them before they did even more damage, hastily shipped them across the Bosphorus. On the Asian shore, they met Muslim warriors and went into battle for the first time. Undisciplined, ill-armed and ignorant of military tactics, they were ambushed, routed and finally slaughtered.

The aristocracy of Europe, meanwhile, were organising their own armies. Germany did not do so. Perhaps the experience of a Crusade of peasants passing through their cities had been more than enough. Neither did Spain; the country had its own problems with the Moors still occupying two-thirds of it. But the noblemen of southern Italy, of Flanders, of Lorraine in the north of France and Provence in the south, each raised a force several thousand strong. The four armies travelled across Europe independently and, in the autumn of that year, met for the first time in Constantinople.

Ravages in the Holy Land
The Islamic horsemen wore flexible chain mail and round helmets, as shown in this illumination from a History of Jerusalem written in the 13th-century AD.

The mounted knights were the Crusaders' main strength. Each man was heavily armoured, with a metal helmet and a coat of chain mail. He carried a sword, a lance and a huge triangular shield covered with leather that protected him from shoulder to knee. His horse, as was necessary to carry such a burden, was large and powerful. Some of the grander and wealthier among them were accompanied by their wives and children. For every one of these knights, there were about seven unmounted men. Some had been serfs in the European estates of the knights and had set out in order to accompany and look after their masters. Others had joined the Crusade quite independently as an act of Christian piety, or simply to see the world. These foot soldiers were mostly armed with pikes and bows. Their duties were to tend the knights, to defend the camps while the knights were away seeking the enemy, and to help in the major construction work when a town or a

castle was besieged. In all, the assembled Crusader army numbered at least thirty thousand men. Some estimates put the figure as at least twice that.

Even at this early stage, there were major disagreements among the princely leaders of the different groups as to who should take command; but at last matters were settled, superficially at least, and the immense host was ferried from Constantinople across the Bosphorus into Asia to tackle the enemy.

A European knight's customary way of fighting was to mount his horse and charge his opponent in an attempt to unseat him with a blow from his lance, and then if that failed, to fight at close quarters, slashing at him with a huge sword. That was not the Islamic way of war. The Turkish warriors were far more lightly equipped than the Crusaders. They wore leather tunics only sparsely covered with iron plates and carried small, round shields. Although they possessed short swords, their favoured weapon was the bow and arrow, which they used with consummate skill, if necessary without dismounting or even halting. Their horses were small, swift animals like those that had carried Muhammad's armies. Their tactics too were very similar. Bands of them would burst upon the massive Crusader columns, discharge a shower of arrows, and disappear before the Crusaders could organise any effective response. If the Crusaders then broke ranks to follow, as likely as not they rode straight into an ambush. Only when the Crusaders' formation was broken up and demoralised did the Turks move in to fight at close quarters.

The first major set battle was fought for the city of Antioch. The Crusaders arrived in front of its walls in the summer of 1097 and settled down to besiege the city. But there was to be no easy quick

Christendom and Islam do battle
A 13th century manuscript from Acre (*above*) shows the Crusaders besieging Antioch. Their heraldic banners and horse-cloths were not, in fact, adopted until later in the campaigns. On open ground, the horsemen fought with lances. A heavily armoured Christian knight (*left*), braced in his saddle, gives a glancing blow to his Muslim opponent who is so unbalanced that his feet fly from his stirrups.

victory. As winter approached, the besiegers ran almost as short of food as the besieged, and they had to send large detachments of their men on journeys of several hundred miles searching for fresh supplies. Plague broke out in their camp, and great numbers of them died. Many deserted and tried to find their way back to Constantinople. Those that remained continued quarrelling among themselves, this time as to who should have the right to pillage the city when it eventually fell. It did not do so until the following year.

No sooner had the Crusaders entered it and begun to sack it than they themselves were besieged by a Muslim army that had come, too late, to rescue the city. But at last the Crusaders forced their enemies to withdraw and they were able to move on towards their goal. They reached Jerusalem a few months later, besieged it and on July 14th, 1099, they took it. They celebrated their victory with the biggest and most brutal massacre so far. Many Muslim citizens – men, women and children – had taken refuge inside the city's main mosque. The Crusaders barricaded it, set fire to it and burnt alive all those within. They treated the synagogues, packed with Jews, in the same way. Christianity had returned to the Holy City.

To maintain their hold on the land, the Crusaders now set up four small states between Constantinople in the north and Egypt to the south, each ruled by a European nobleman. Within these tiny kingdoms military power was centred on a series of castles. Krak des Chevaliers, standing on a hill in Syria some twenty miles from the coast, is today one of the most perfectly preserved. It is surrounded by two concentric rings of walls, separated by a moat. Each is studded with projecting towers from which defenders could fire arrows at anyone who tried to scale the walls with ladders or to undermine them. The only entrance to the castle is over a drawbridge and through a heavily fortified gateway. If attackers managed to get past these two barriers, they were faced with a long dark passage up which they had to fight. After eighty yards, it suddenly makes a hair-pin bend behind which fresh forces of defenders could be waiting unseen. The stone roof above is pierced with holes through which the defenders could hurl boulders, fire arrows, and tip burning pitch and oil. If the attackers survived all this, they then had to face the massed knights waiting to do battle with them in the central courtyard. No one, in the castle's history, ever managed to get so far. The castle, in fact, was virtually impregnable against direct attack.

Like most Crusaders' castles, Krak was far more than a military strong point. It was an administrative centre to which the people brought their taxes; a monastery where men of God, armed though they might be in the service of Christianity, could lead devout lives; a hospice where pilgrims on their way to the Holy Land could find safety and shelter; and a palace where the European nobleman ruling the province could dwell in suitable magnificence. At the height of its power, it was home to some four thousand people. The commander had his quarters in a tall tower which overlooked all the castle and far across the country beyond. Serving him were a hundred or so Knights Hospitallers of St John, who had taken monastic vows of poverty, chastity and obedience to their Commander and who had made it their special task to protect Christian pilgrims and maintain public order. They lived in lodgings in the courtyard at the foot of the commander's tower alongside the refectory and chapel where they ate and prayed together. There were also a great number of locally recruited general helpers and labourers.

Several hundred horses for the knights were stabled in an immense vaulted hall just inside the main gate. An aqueduct running from the

The siege of Jerusalem
The 15th-century artist who provided this illustration for a history of the Crusades had not, it seems, visited the Holy Land, for he provided Jerusalem with very Gothic churches. The techniques used in the siege, however, he showed more accurately, for they were much the same as those practised in European wars – scaling-ladders, siege-towers on wheels, huge timber screens, also mounted on wheels, behind which the besiegers could shelter, and fires built from bundles of faggots heaped around the walls.

Krak des Chevaliers
The main entrance (*1*) led into a long roofed corridor with a hairpin bend (*2*). This opened through gateways (*3* & *4*) into the central courtyard where the knights had their lodgings, cloisters and refectory (*5*). The commander had his own quarters in a tower (*6*) overlooking the whole castle and the countryside beyond. In times of peace, water was supplied by an aqueduct (*7*).

A deep ditch lies between the nearest spur (*right*) and the castle. It was from this side that the Krak was most vulnerable. (*Below*) The hairpin bend in the corridor between the entrance and the central court.

Islamic irrigation
In the Syrian city of Hama on the River Orontes, huge wheels driven by the flow of the river lift water into conduits, along which it flows to gardens and fields, as they have done for a thousand years. Such skills in irrigation enabled the people to cultivate groves of oranges (*below*), and other crops, on land that was otherwise too parched to be productive.

nearby hillside kept the moat filled. Since that could easily be blocked during a siege, rainwater was also carefully collected and stored in huge cisterns cut in the rock. And in the lower part of the castle lay huge store rooms stacked with food and other supplies sufficient to sustain the castle through months or even years of siege.

The land in which the Crusaders had settled astonished them by its richness and fertility. The techniques of irrigation that the Moors had introduced into Spain were used here on a spectacular scale. Water distribution systems dating from these times are still functioning in the Syrian city of Hama. Huge water-wheels, one seventy-five feet high, creak and whirr as they turn continuously day and night driven by the Orontes river. Each blade of these gigantic cog-wheels is a trough which, as it dips into the river and is pushed forward by the current, fills with water. As the wheel revolves, so the troughs rise until they reach the highest point and begin to descend again, when the water tips out of them into a high aqueduct. From there it flows down to gardens throughout the town.

With such techniques, the people in the lands around the castles grew a great range of crops, many of which were quite unfamiliar to the Crusaders. There were orange trees which, they noted wonderingly, fruited at the same time as they flowered. Figs, grapes, pomegranates, melons and cucumbers all grew prolifically. Fields were filled with cotton bushes with fluffy white seeds from which the people wove a soft thin cloth. They cultivated a tall cane which they had imported from their territories farther east in India. They squeezed juice from this, concentrated it by boiling and so produced sugar, a substance new to the Crusaders who, in their home lands, had only honey as a sweetener.

Not surprisingly the Crusaders, settling down in castles where they might well spend most of their lives, began to adopt the customs of the people around them, as many colonial administrators have done both before and since. They developed a taste for the food of the country. Many of the dishes were flavoured with spices that few of the Europeans had tasted before – cloves and nutmeg and cinnamon. Cane sugar was used to make not only delicious, sticky sweetmeats, but also drinks flavoured with orange and lime – both *sherbet* and *sorbet* are Arab words.

The Crusaders and their families also took to the local standards of hygiene. Men coming fresh from Europe to reinforce the garrisons were shocked to discover that some of the Christian ladies who had lived for some time in the East not only perfumed themselves with attar of roses, a fragrant oil made by boiling rose petals, but also patronised the local public baths. They even dusted their skins with a powder the Arabs called *talcum*.

In dress too, the Crusaders changed. The women adopted the Islamic style of long clinging garments, and occasionally the veil; and the men started to wear silk cloaks embroidered with thread of silver and gold, and white tunics over their armour which kept them cool. A few even took to turbans. Beards in Europe were not fashionable at this time, but cutting facial hair was a disgrace to a Muslim and soon the Crusaders were also growing their beards.

Even though these European enclaves absorbed so much from their surroundings, the land in which they stood was still an alien one, and its people were still fundamentally hostile to their new rulers. The castles always had to be prepared for an attack: raiding parties of Muslim horsemen might appear from nowhere. Even if they were not strong enough to attack the castle, they might fell orchards and destroy crops around it simply to deny food to the garrison. When large-scale forces did arrive and laid siege to a castle, the effect on the countryside was devastating. Thousands of Muslim soldiers camped around the walls and prepared to stay for weeks or months. They attacked the castle with rams, huge tree trunks, their ends sheathed in iron, with which teams of men, under a protective roof, attempted to batter a hole in the base of the castle wall. Mangonels, long timbers pivoted near the middle and heavily weighted at the shorter end, were used to catapult boulders over castle walls. Sometimes the missiles were the severed heads of Christians who had been killed during a skirmish or, worse, taken prisoner. The attackers might build siege towers in full sight of those manning the castle walls but beyond the range of their arrows. These towers had to be taller than the walls, and could stand sixty or seventy feet high. Building them took many weeks, but when one was at last finished, its front flank was hung with animal skins and wet mats to prevent it being set on fire by missiles of burning pitch, and it was slowly pushed towards the wall on wheels or rollers. When it was close enough, the men within it lowered a bridge and rushed across to battle hand-to-hand with the castle's defenders.

Every one of these devices demanded huge quantities of timber for its construction; and the great armies needed tons of wood for their fires. So the inauguration of a siege meant inevitably that all the trees for miles around would be felled. So the land which had astonished the newly-arrived Christians by its productivity was, year by year, laid waste.

Muslim armies came from surrounding states, from Damascus and Mosul, Mesopotamia and Egypt to attack the Crusader States. Castles were taken by storm or starved into submission. Some changed hands several times between Christians and Muslims, on each occasion the victors repairing the damage they themselves had inflicted and then improving the defences with newly devised features; so the castles

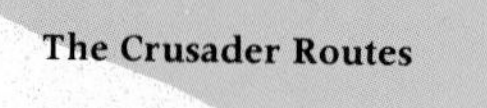

became composites of Christian and Muslim ingenuities of fortification. The balance of power swayed back and forth, but the Christians could only maintain their position if they had regular reinforcements of fighting men. The most direct route for those to come was along the length of the Mediterranean from the ports of Sicily and Italy. To protect that route, the Knights Hospitallers of St John together with the Knights Templar, a similar order, took possession of the islands of Rhodes and Cyprus. The clergy in Europe called repeatedly for more armies to go to the Holy Land. Contingents were raised in countries, including England and Germany, that had not taken part in the first expedition; and six more Crusades were launched. But eventually, after a hundred and fifty years, Europe began to lose its zeal for this never-ending war in a distant land.

Krak des Chevaliers never fell to a direct assault. It was attacked many times but its superb bastions remained unbreached. By 1271, however, the Crusaders were so few in number that Krak had less than a quarter of the men it needed to defend its walls properly. Once more a Muslim army laid siege to it. This time, after only a month, the dispirited defenders surrendered it in exchange for a safe passage down to Tripoli on the coast twenty miles away. There they took ship for Europe. Over the next twenty years, the last of the Crusaders straggled back home. Western Europe's occupation of the eastern shores of the Sea had come to an end.

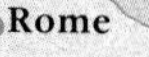

EASTERN ANIMALS MOVE WEST

The returning Crusaders brought the Orient to northern Europe just as the Moors had taken it to Spain. Some had with them fine steel blades which they described as 'damascened', since they had been patterned and inlaid with gold and silver by the craftsmen of Damascus. Their ladies had damask silk which came from the same city, as well as muslin, a delicate cotton fabric from Mosul. Once settled back home, they built new houses or modernised their castles, and constructed special rooms in which to continue their new habit of bathing. They even installed small platforms piled with cushions of the kind the Arabs called a *sofa*. Many of the marvellously wrought objects they had collected as mementoes of their eastern campaigns – pile carpets, gilded glassware, bronze lamps – were presented to the local church, where they were kept in altars, in spite of the fact – or perhaps more accurately, in ignorance of it – that some were inscribed with indecipherable texts praising Allah.

In addition to all this baggage, they also brought an animal, the black rat. Within a generation it was to have as dramatic an effect on medieval Europe as the horse had had on Rome. Its original home was in India, but it had long since established itself in the Middle East, and had spread from there into western Europe. Being a semi-tropical animal in origin, it relished warmth and could not withstand a really cold winter out in the open. So it did not spread far into the woodlands of the north. But it did find the shelter it needed in the homes of men. There it also discovered abundant food, in the store rooms and in the refuse the human beings littered around their dwellings. So the rat became man's ineradicable companion, nesting in the roofs of his houses, feasting in his larders, swarming in the rigging of his ships. By the end of the thirteenth century, it had appeared in European ports and cities. There it found a ready home, for the more thickly furred, hardier brown rat which, five hundred years later, was to displace it over much of the continent, had not yet spread into Europe from its native grasslands in central Asia.

In many places the black rat increased to plague proportions. Towns had to employ special rat-catchers, and the tale of the piper in the German town of Hamelin who first charmed all the rats of the town to their deaths in the river and then, when the town refused to pay his fee, enticed away all the children, dates from soon after this time.

Although the black rat caused very considerable damage, it did no direct physical harm to man. But it was subject to a particular disease. It carried a bacillus in its blood which produced a lethal septicaemia; and it was also infected with fleas. When an infected rat died, its fleas, with their guts full of its blood, hopped away to find another host. Usually it was another rat. Eventually, it was a human being. It seems that the Crusaders brought back such infected rats in their baggage.

The Great Pestilence broke out in Europe in 1347. It first appeared in Sicily but soon it was raging all over the continent. Accounts of it vary, and it may be that several clinically distinct diseases were involved, but one of them, it seems certain, was the bubonic plague that was carried by the fleas of the black rat. The bodies of those who caught it became covered with 'bubos', swellings and carbuncles particularly in the groin and the arm-pits. Their breath became foul, they vomited blood and then died within a day or so, sometimes within a few hours.

There had been many outbreaks of epidemic diseases before, but this time the crowded, insanitary conditions of the growing towns, and the recent arrival of infected black rats produced a catastrophe on a scale that was unparalleled. The horrifying speed with which the disease killed, and the rate at which it spread, terrified Europe. Families among whom it appeared were walled up by their neighbours

The carrier of the plague
The black rat has larger ears and a much longer tail than the brown rat (which today is much more common in most of Europe, and which sometimes is almost as black in colour). It is a much more agile climber and, instead of making its home in burrows as the brown rat does, prefers to nest in roofs and lofts. It has also, in the past, carried in its blood the deadly bacillus of the plague.

in an attempt to prevent it spreading further. But the walls were no barriers to rats. Some people deliberately inhaled the stench from sewers for several hours a day, hoping that such foul gases might eradicate whatever it was that caused the infection. The bodies of the dead were hastily piled into pits and buried as quickly as possible. But there was no way of stopping it. Within three years of the first appearance of the Great Pestilence, one person in three in Europe had died of it.

The population of the continent before the epidemic struck was about fifty million, roughly the same as it had been at the time of the collapse of the Roman Empire. Although a great deal of land had been cleared for cultivation since that time, and towns were growing, there were still vast tracts of woodland that were largely unaffected by man's presence. Beavers were still abundant in the rivers; bears and elks, ibex and lynx were still regularly hunted. Even bison and the ancient wild cattle could still be found.

The population grew rapidly after the disappearance of the Pestilence, and now the pace at which man cut down the woodlands began to increase sharply. As more land was cleared for farms and more food produced, so more children survived, families became bigger and, in turn, demanded more land on which to farm. And now a new claimant appeared. With the expansion of trade, commercial ways of exploiting the land became evident and, in Spain, the forests were sacrificed on a devastating scale for an animal that was bringing great wealth to its owners, the Merino sheep.

The origin of this breed is not wholly certain, but it seems likely that it first appeared in North Africa and that it was brought into Moorish Spain during the twelfth century by the Beni-Merino, a nomadic tribe of sheep-keeping Moors from the Atlas Mountains. The wool it produced was particularly long and fine, and it could be spun and woven into a fabric so warm and waterproof that it began to replace the linen made from flax which, until now, had been the main material used for clothes.

The Spanish kept their Merinos during the winter on the southern lowland pastures; but when the hot summer came, the grass there dried and withered, so the sheep had to be driven five hundred miles north up into the mountains where each spring the melting snows uncovered freshly sprouting pastures. There the sheep were grazed through the summer months until the autumn came once more and they were driven south, back to the lowlands.

The trackways, called cañadas, along which the sheep were driven, had long been established, but as the demand for Merino wool and the cloth made from it grew all over Europe, so greater and greater numbers of the breed were kept, and the demands on the land through

The triumph of death
This 15th-century painting from Palermo shows Death, riding his skeletal horse, armed with his scythe, striking down young and old, rich and poor, saint and sinner, as he did during the Great Plague when a third of the population of Europe died within three years.

which they passed on their month-long journeys increased. The aristocrats, to whom most of these flocks belonged, formed themselves into a Society called the Mesta to represent their interests at Court and to lobby for their rights. The King of Spain hardly needed encouragement to look upon the members of the Mesta with favour, for he received a tax from every sheep they owned and every bale of wool they exported. Eventually, he himself became the owner of huge flocks, and in 1476 he was made the Grand Master of the Mesta. The Society was now all-powerful.

By the end of the fifteenth century, the last of the Moors were driven from Spain and even more of the southern pastures, in Andalucia and Estramadura, became available to sheep. By 1526, there were some three and a half million Merino sheep in the country and they constituted a major element in the State's economy. The King,

The long march of the Spanish Merinos
In the 15th century, the land between the winter grazing in the lowlands of southern Spain, and the summer pastures in the mountains was largely occupied by farms and forests. Today the forests have mostly been felled, and large industrial cities have developed. The sheep still make the journey, but they travel for the first part in trucks by rail. They then trot through the streets of the railhead towns (*left*) bringing the traffic to a halt, and continue upwards for several days until at last (*above*) they reach the grass of the bare uplands, newly revealed by the melting snows.

anxious to preserve the Spanish monopoly of the breed, and hence his income, made it illegal for live Merinos to be exported, and passed stringent laws to aid the Mesta. The flocks, as they moved between mountain and plain, had to feed. The wide cañadas had hitherto provided all the grazing that was necessary, but as the flocks grew bigger and bigger, they were not enough. Farmers with land beside the trackways were forbidden to fence their fields to keep out the hungry sheep. More and more of the forests were felled to widen the trackways, and land that had already been cleared was requisitioned by the State on behalf of the Mesta. Anyone refusing to give up his land to the sheep when asked to do so was subject to the death penalty. Eventually, the cañadas became two hundred and fifty feet wide, huge ribbons of land stretching over half the width of the entire country. Up in the mountains, more of the forest was felled so that the bare grass-covered slopes extended downwards to lower altitudes. In the lowlands, the shepherds were given permission to cut the young shoots from trees to feed the sheep during the winter. By the middle of the sixteenth century, the forests that had once covered most of central Spain had largely gone. Over great areas, the land was totally treeless and its topsoil had washed away. The agriculture of Spain has not recovered to this day.

The Spanish king also owned estates in Italy. He sent Merino sheep there as well and instituted the same grazing regime. The effect on the Italian forests was very much the same, but here the trees were also being cut down by an even more voracious industry, ship-building.

THE DESTRUCTION OF THE FORESTS

By the middle of the fifteenth century, after the end of the Crusades, Venice had emerged as the greatest Christian sea-power in the Mediterranean. She owed that position to her pre-eminence in trade. Ideally placed, mid-way between the source of spices and other Oriental goods in the Near East and the increasingly affluent lands of western Europe, she managed to make trading agreements with both the Turks and the Egyptians. Her ships collected goods from Constantinople and Alexandria and carried them to France and Spain and beyond the Gibraltar Straits to England, Flanders and the rest of northern Europe. Her ruler, the Doge, ever since the eleventh century, had acknowledged her debt to the sea by annually performing the *sposalizio del mar*, a splendid ceremonial during which he was rowed out into the centre of the lagoon and there, representing the State, was symbolically wedded with the waters. To protect their trade routes, the Venetians had acquired not only a long strip of the coast of Dalmatia on the eastern side of the Adriatic but also extensive territories to the south on the Greek mainland, many of the smaller Aegean islands and Crete. She also took over Cyprus from the Crusaders and so had a forward base with excellent harbours from which to trade with Egypt and the Levant.

At the beginning of her expansion, it must have seemed that she had ample supplies of timber for ship-building, for her forests stretched from the margin of the lagoon in which the city stood to her farthest inland frontier on the flanks of the Alps. They contained all the different kinds of trees that were required: spruce and fir for masts, larch for planking, elm for capstans, walnut for rudders and beech for oars. Most important of all, there were oaks. Their straight trunks were needed for keels and deck beams, their massive curved boughs for ribs. Even while the Crusaders were still fighting, Arab traders came to Venice to try to buy timber, and the Pope had to intervene and forbid the merchants of the Serene Republic to give support to Christendom's enemies, even though there might be considerable profit in it.

A Venetian galley
Venice still today holds an annual regatta for traditional craft. In the 13th century, the event took place out in the lagoon. During the celebrations, important ceremonials were enacted, as well as a long series of races which served as a means of training and selecting oarsmen to row in the galleys of the State.

Venice had two kinds of ships in her fleet. The most numerous were the tall stately merchantmen, broad in the beam and with capacious holds, driven by sails. She had about three hundred of these, and between them they carried the bulk of the Republic's trade. Thirty were particularly large, weighing over 240 tons, and these made the long international voyages, taking silk, woollen cloth, glass and metal-work from Venice, and bringing back cotton from Syria, wine from Crete, slaves from the Black Sea, and grain and olive oil from ports all around the Mediterranean.

But her crack, élite ships were the galleys. They were, by contrast, slim and low. Although they might have sails, in which case they were

more correctly described as galliasses, they also carried about a hundred and fifty oarsmen. They manned the huge oars, and could drive the ship forward when she was becalmed, or take her in and out of harbour even when the wind was so unfavourable that sailing ships were unable to leave or enter. These oarsmen were all free men, not slaves as they would become later (and already were elsewhere), and they were armed, so they were well able to defend the galley from attacks by pirates. Being safe, predictable and fast, the galleys were ideally suited to carrying precious cargo and they had the monopoly of Venice's lucrative spice trade. They operated on regular schedules all around the Mediterranean. One service ran to the Barbary coast of

North Africa by way of Sicily. Another went through the Gibraltar Straits to Lisbon and on to England and Flanders. For those that could afford to use them, they were the best and most prestigious way of travelling, the very symbol of Venice's pre-eminence among the sea-going nations of Europe.

Galleys also served her as her main warships. They fought by ramming an enemy so that their crews could board and fight their opponents hand-to-hand. She built these impressive ships, for both warfare and commerce, in the Arsenal, the state shipyard.

By the end of the fifteenth century, the Venetian merchants were having considerable difficulty in obtaining the tall sailing ships they needed for their bulk trade. The civilian shipyards that provided them could not find the timber they needed in the Venetian territory inland, and the merchants were therefore having to buy vessels from Dalmatia, in particular from the city of Ragusa. This did not suit the Venetian authorities, who wanted to protect the city's own industry. They therefore imposed special levies on vessels imported from abroad, and passed regulations forbidding foreign-built ships to fly the Venetian flag, even if they were Venetian-owned.

The Arsenal had to be kept supplied at all costs. The Republic could not willingly become dependent on foreign powers for the supply of her warships; so officials of the Arsenal scoured the Venetian territories on the Italian mainland in search of trees. They identified the few individual trees standing outside the forests that were big enough to provide the size of timber needed and marked them as property of the State. They laid claim to the surviving areas of forest and prohibited all felling in them. Guards were stationed to make sure that the regulations were obeyed. Indeed, Venice can be credited with having instituted the world's first carefully planned and administered programme of forest conservation. But it was too late.

Venice's fighting fleet was essential to her survival as a trading power, for the war between Christendom and Islam had by no means ended when the Crusader armies returned from the Holy Land. It was now being fought at sea. In the east the Turks, having taken Constantinople in 1493, were extending their power across the Aegean. They had occupied many of the smaller Greek islands, driving out the Venetians in the process. In 1522, they landed on Rhodes with an army of two hundred thousand men. The Knights of St John, who had been established there for over two hundred years, had fortified their city around the harbour so skilfully that their garrison of only five thousand men were able to keep the vastly superior Turkish army at bay for six months. But eventually they had to surrender, and were allowed to sail back towards their home countries.

At the west end of the Sea the Moors, who had been driven out of

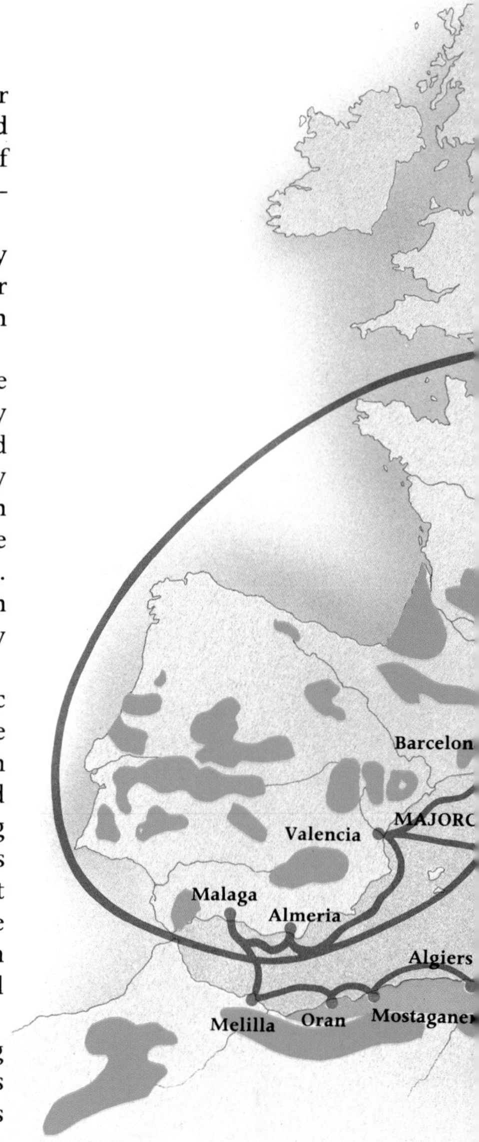

Spain, had settled on the North African coast around Algiers. Their galleys, manned largely by slaves, were harassing the Christian fleets, and they became known and feared throughout Europe as the Barbary corsairs. They sailed along the coast of southern Spain, which had been their home for seven hundred years and where they could still find many to give them help, and plundered the rich cities. They raided all along the western coast of Italy, pillaging and capturing men to enslave and shackle to the oar benches of their galleys. Corsica,

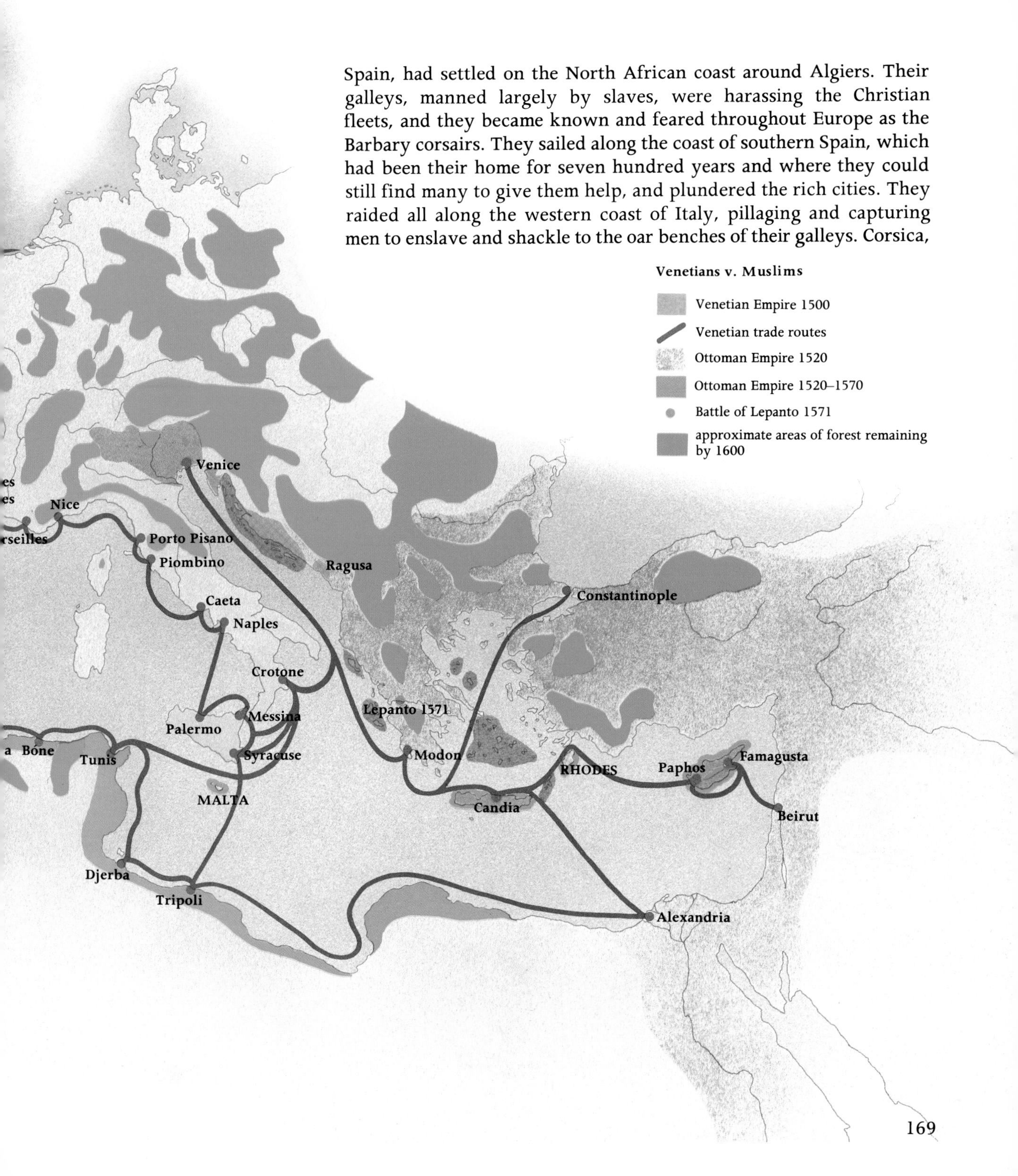

Sardinia, the Balearic islands, even Naples, Rome and Genoa were all attacked. Nowhere along the coasts was safe from them.

In 1533, the most daring and successful of the corsair captains, known to Muslims as *Kheir-ed-Din*, Protector of the Faith, and to Christians as Barbarossa, Red Beard, was summoned by the Sultan of Constantinople and given the task of reorganising and enlarging the Turkish fleet. Under his command, an immense force of over a hundred galleys and many smaller supporting vessels began to wreak havoc in the eastern half of the Sea. They captured and devastated

The last battle of the galleys
The Battle of Lepanto, which brought Islam's dominance of Mediterranean waters to an end in 1571, provided a popular subject for patriotic Venetian painters for a century or so after it was fought. But it was the last in which Venetian ships were to play a significant part.

Crete. Year by year, their power grew. In 1563, they ventured west along the Sea and once again attacked the Knights of St John who, by now, had re-established themselves on the island of Malta. A fleet of over two hundred vessels, with over forty thousand men aboard, fought for the island for four months. But once again, the Knights had built their fortifications with consummate skill, and they defended themselves with great valour and heroism. This time, the Turks had to retreat. But they were by no means daunted. Eight years later, in 1571, they attacked Cyprus, the last of Venice's outposts in the eastern part of the Sea.

Venice made urgent pleas for help and at last the Christian nations managed to settle their differences and combined to launch a counter blow. Venice provided half the fleet, but ships also were supplied by the Pope, by Genoa, Spain and Malta. They met the Turkish fleet in Greek waters at Lepanto. The Christians had eight large galleys, two hundred smaller ones and many other supporting ships, making some two hundred and fifty vessels in all. The Turkish fleet was slightly larger. In a narrow strait, the great galleys, driven by their huge oars with three men straining at each, smashed into one another. The fighting on the decks raged back and forth. But eventually the Turkish flagship was taken. Its huge pennant, inscribed twenty-six thousand times with the name of Allah, was hauled down. The Muslim admiral was captured and executed, there and then, on the deck of his ship and his severed head hoisted on a pike to be displayed to both victors and vanquished. It was one of the most crucial sea battles in Mediterranean history and it marked the end of Turkish dominance as a naval power.

The cost was immense. Twenty thousand Turks were killed or captured. Eight thousand Christians lost their lives and some sixteen thousand were seriously wounded. But the battle was not only to be measured in men's lives. It also had to be paid for in trees. The flagship of the Christian fleet was a galley, *El Real*. We know exactly how much timber was needed for her construction because recently an exact replica has been built. Fifty beech trees were required for her oars, three hundred pines and firs for her planks and spars, and over three hundred mature oaks for the timbers of her hull. Altogether the fleets that fought that day had necessitated the felling of over a quarter of a million mature trees.

Lepanto was the last major battle in which galleys played an important part. The introduction of cannon on to ships was changing the whole character of naval warfare. Improvements in rigging were making sailing ships almost as manoeuvrable as those with oars and they required many fewer men in their crews. The merchant galley, too, had become obsolete, since the Portuguese had found their way

El Real, last of the great galleys
Six hundred and fifty trees were felled in order to build her.

around Africa to India and had tapped the spice trade at its source, robbing the Venetian galleys of the most important element in their cargo. But in addition to all this, the building of galleys in the Arsenal at Venice had come to an end because the Mediterranean forests were now almost exhausted of timber. Venice was no longer the queen of the western seas. The flag ship of the Christian fleet at Lepanto was not hers. *El Real* had been built in Barcelona in northern Spain. Slowly, over the next few years, the centre of ship-building moved out of the Mediterranean, farther and farther north. Eventually it came to the Baltic, where the yards were able to draw on the vast and still largely unexploited forests of northern Europe. The Mediterranean countries could no longer sustain their own fleets.

After the forests had been destroyed, the fate of the stricken lands was sealed by the goat. This animal had been domesticated around the same time as the sheep, but in some respects it is less useful to man. Its coat is coarser and not as easy to spin as wool, its meat is tougher than mutton, and it produces very little fat. But it has two great qualities. It produces a great deal of milk for its size; and, most useful of all, it does not demand grass as sheep do, but will eat anything green, no matter how fibrous or thorny. What is more, it will even climb trees to get it, balancing itself in the most surprising fashion on boughs in order to reach the last leaf. So it flourishes in arid country where cattle or sheep would starve.

Goats had long since become the commonest domestic animal along the African shore. There, the herds grazed among the fallen columns of the ruined Roman cities. The Turks, who had never practised irrigation with the skill and dedication of the Arabs, had brought great numbers of goats with them into their newly acquired territories in the eastern Mediterranean, and now the herds spread all over the Levant and Greece. Once goats are established, the land stands little chance of recovering its trees and regenerating its topsoil. The goats consume every seedling that sprouts and every leaf that unfurls. The people who keep them can hardly be expected to banish such a useful creature that can produce so much from so little. So the land remains barren.

The woodlands and forests that had once ringed the entire Mediterranean and provided a home for a rich population of animals had now largely gone. They survived only in places for which no soldier bothered to fight, where no forester found it possible to haul out timber, and no farmer thought it worthwhile to make a field. Now the maquis and the garrigue, which had once grown only on the rockiest and most impoverished stretches of the coasts, spread all around the Sea and dominated what were once among the most fertile lands in the world.

PART FOUR

STRANGERS IN THE GARDEN

Goats browsing trees in Morocco.

MARINE INVADERS FROM THE EAST

Although, by the sixteenth century, human beings had radically changed the face of the lands around the Mediterranean, their impact on life in its waters had been minimal. True, people had begun to fish as soon as they arrived on its shores about a million years ago, and they had continued to do so with increasing skill and intensity as the centuries passed. The Cretans, the Greeks and the Romans all left vivid evidence, drawn on vases, portrayed in mosaic and painted on walls, of how they did so and what they caught. But fishermen remained so few in number, and their methods were so simple, that they did not change the character of the marine world. It remained a slightly impoverished version of the Atlantic community from which it had originated five and a half million years earlier. The surface current flowing into the Sea through the Gibraltar Straits continued to bring in fresh planktonic recruits, and the heavy, salty bottom current continued to flow out westwards, taking with it a significant quantity of nutriment.

A hundred miles south of the eastern end of the Mediterranean, however, lie very different waters. The Red Sea is tropical. It is the only true sea in the world that is even saltier than the Mediterranean, for not only is it closed at one end, but the sun shines on it with even greater intensity, evaporating its waters even more swiftly. Furthermore, unlike the Mediterranean, no river flows into it, and no rain falls on it to dilute its saltiness. Coral reefs, unknown in the Mediterranean, fringe its shores; and its fish, which swarm in great numbers and variety, are derived not from the Atlantic, but from the vast and diverse population of the Indian Ocean. Very few species have spread from the Indian Ocean round the Cape of Good Hope, and into the chilly waters of the North Atlantic; and very few if any of these have found their way through the Straits of Gibraltar.

The narrow neck of land that separates these two very different marine worlds is flat and sandy, and formed largely of sediments brought down by the Nile. At its highest, it is no more than seventy feet above mean sea-level. Both the Red Sea and the Mediterranean have at different times flooded part of the way across it, as semi-fossil shells and marine deposits testify; but at no time did the waters from the two sides meet. In the middle of the isthmus lies the extremely salty Lake Timsah, and south of that is a group of briny lagoons known as the Great Bitter Lakes.

The rulers of ancient Egypt saw little advantage in linking the two seas, but they considered that much could be gained if the ships

The link between two seas
A satellite view of the Suez Canal cutting through the far corner of the huge triangular delta of the Nile, and running across to Lake Timsah and the Red Sea.

plying up and down the Nile were able to sail directly into the Red Sea and so reach the markets of Arabia and the Horn of Africa. So at the beginning of the sixth century BC, the Pharaoh Necho II started to excavate a linking channel. At that time, a branch of the Lower Nile ran eastwards from where the city of Cairo now stands, to empty into the Mediterranean at the town of Pelusium at the easternmost corner of the delta. Necho's plan was to dig a canal from half-way along this branch, due east of Lake Timsah, and then south through the Bitter Lakes to reach the most northerly tip of the Red Sea at Suez. The labour involved was immense. The Greek historian, Herodotus, stated that one hundred thousand people were killed as the work proceeded. Eventually, the Pharaoh was warned by an oracle that the human cost was too high, and accordingly he ordered the work to be halted.

In 520 BC, however, Darius, King of the Persians, invaded Egypt and he commanded that the work should be started again. He apparently saw it not merely as a useful trade route but also as a monument to his own glory, for he set up at least four stone tablets along its course, declaring that he, as the conqueror of Egypt, had decreed that the canal should be dug. It was, certainly, a spectacular achievement. Strabo, the Greek geographer who lived at the beginning of the Christian era, described the completed waterway as being wide enough to accommodate abreast two triremes – war galleys with three tiers of oars on each side – so it was probably at least a hundred feet across. The journey from the Nile to the Red Sea took, he says, about four days. The canal was, however, very difficult to maintain. Sand continually drifted into it, silting it up. It was cleaned out and repaired several times by the Romans during their rule of the country, and again by the Arabs when they annexed Egypt in the eighth century AD. Eventually, the Arabs decided it was more trouble than it was worth, and they filled it in altogether.

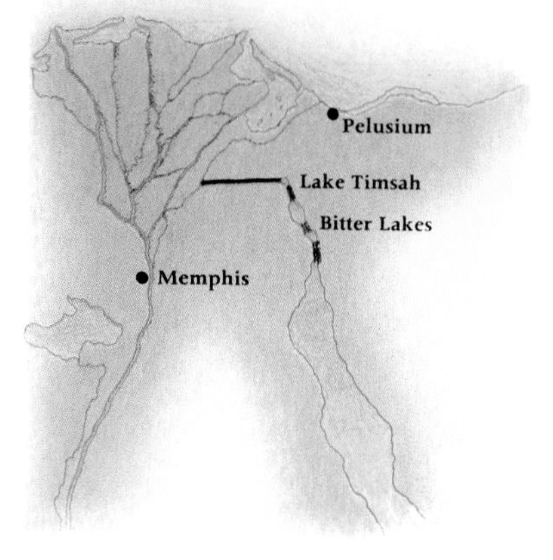

The Pharaoh's Canal
Like its successor, built over two thousand years later, Necho's canal linked pre-existing lakes. It did not open directly into the Mediterranean but joined the main eastern branch of the Nile, along which shipping from the Red Sea could then sail to reach the Mediterranean at Pelusium.

A thousand years later, the faint remains of it in the desert were personally inspected by Napoleon when he took over Egypt in 1798. The advantages of connecting the Mediterranean with the Red Sea were very clear to him. He imagined the ports of southern France linked through her new Egyptian colony to India and the Far East, and he ordered his engineers to examine the possibility of bringing this about. They surveyed the isthmus and reported that, according to their calculations, the Red Sea was thirty feet higher than the Mediterranean and that if the Indian Ocean were not to flood catastrophically northwards, locks would be needed. So Napoleon abandoned the idea.

Fifty years later, it was resurrected by another Frenchman, Ferdinand de Lesseps, who was the French consul in Cairo. He concluded that Napoleon's survey was wrong and argued that a canal could and should be built. With extraordinary persistence he convinced enough people that this was so, and in 1859 the first spadeful was shifted. Initially the work was done by thousands of locally recruited labourers using picks and baskets. De Lesseps, like Necho, planned to link the existing lakes, so that less than half the total length had to be excavated in dry land. After some time, it was discovered that dredging was cheaper and quicker and the cuttings were artificially flooded so that such machinery could be floated in. It still took twice as long to build, and cost twice as much, as de Lesseps had told his backers; but eventually it was opened with huge celebrations in 1869.

De Lesseps's surveyors were proved right, and Napoleon's wrong. There was only the slightest difference in level between the two seas, the Red Sea being marginally higher, and ships were able, without the impediment of locks, to sail straight through the canal. It was one hundred and five miles long, twenty-six feet deep and seventy-two feet wide at the bottom. Since that time, it has been deepened to forty feet and widened to one hundred and seventy-nine feet. Ships have to move through it very slowly so that their wash does not erode the banks unduly. It takes about fourteen hours. Before it was built, the voyage from London to Bombay was around 10,700 miles long. The canal shortened that to a mere 6,270.

The Suez Canal revolutionised life in the eastern end of the Mediterranean. The sea route to the Orient round southern Africa found by the Portuguese, and Columbus's crossing of the Atlantic and discovery of the Americas, had drawn shipping away from the Mediterranean and turned the Sea into something of a backwater, a cul-de-sac that led nowhere except to the impoverished lands of Turkey and the Levant. Suddenly it became once again one of the most important travel routes in the world. Today, fifteen per cent of all international shipping sails along it.

Not only ships travel through the Canal. So do marine animals. It is very unlikely that any fish had made the journey along the Pharaoh's canal, for that would have involved moving from the extremely salty Red Sea into the fresh water of the Lower Nile, and then to salt water again in the Mediterranean. Few marine organisms would be able to tolerate such severe physiological changes. But de Lessep's canal, untrammelled by locks, flowed directly from sea to sea: in theory a fish could swim from one to the other without ever leaving salt water.

Even so, the journey would not be an easy one. The fish would have to travel along a narrow channel perpetually disturbed by convoys of immense ships which churn the turbid waters with their propellors, then enter the Bitter Lakes which are even saltier than the Red Sea, and maintain progress up the channel for another fifty miles before

eventually it arrived in water that is considerably colder than its original home. It is doubtful if any individual fish ever made such a journey. The transit could, however, be achieved if fish could establish permanent colonies along the Canal and in the Lakes. Some species, like the brilliantly coloured fish that throng the coral reefs of the Red Sea reefs, have never managed to do this. No coral is allowed to grow along the Canal and, even if such fish were able to reach the Mediterranean, there are no reefs there to provide them with the kind of homes they need, as the waters are too cold for reef corals to grow. But small, bottom-living fish from the Red Sea, that can tolerate muddy water and frequent disturbance, and are not too particular about what they eat, have managed to establish themselves as permanent residents in the Canal.

The first of them was recognised in the Mediterranean in 1902, thirty-three years after the opening of the Canal. Today, at least forty-one different species of Red Sea fish are living there, as well as an even greater number of crabs, molluscs and other marine invertebrates. Organisms that have made this journey are now called Lessepsian migrants by ecologists, in honour of the man who gave them their opportunity.

Having reached a new sea, many of them flourished greatly. Now they are a significant element in the fisheries of the eastern Mediterranean. Not only do they constitute about twelve per cent of all the species of fish in this part of the Sea, but they make up sixteen per cent in weight of the catches brought back to ports in the Levant. As time has passed, they have spread westwards from the mouth of the Canal until now some are found as far away as Tunis and Sicily.

Surprisingly, this great variety and sheer quantity of immigrant fish does not seem to have had a damaging effect on the existing Mediterranean fauna. Often when an alien species is introduced into an environment, it wins its position by ousting some of the original inhabitants. But this has not happened in the eastern Mediterranean. As far as can be discovered, no native species has disappeared, nor even become rarer, as a consequence of the spread of the Lessepsian migrants. It appears that the waters of this part of the Mediterranean were biologically under-exploited, that there was more food and space available than the existing inhabitants could use. This, in truth, is something of a mystery. Its explanation may be connected with the fact that the waters of the Levant are a long way from the natural Atlantic entrance to the Sea and are still poor in fish species, so that the hardy Red Sea fish, accustomed to intense competition in their crowded native waters and willing to eat a wide variety of food, have been able to find ecological niches that were still unoccupied.

It is also significant that the traffic of organisms along the Canal has been almost entirely one way. Very few Mediterranean species have arrived in the Red Sea. This may be partly because the currents run from south to north along the Canal, partly because the salinity of the Red Sea is closer to that found in the central stretches of the Canal and in the Bitter Lakes, so that migrants do not have to make such a severe initial change in their physiology. It is also likely that the Red Sea is such a dense biological community that few of the relatively small number of species which live in the Mediterranean have been able to edge their way into it.

Building de Lesseps's Canal
Initially the work was done by a vast army of labourers (*above*). Only later were dredgers introduced (*left*).

TERRESTRIAL INVADERS FROM THE WEST

Alien plants had been coming to the lands of the Mediterranean from the east ever since the Arabs started bringing them in the seventh century. But at the end of the fifteenth century Columbus's discovery of the Americas launched another botanical invasion. Indeed, one of his motives in crossing the Atlantic had been to discover a new way to the Indies to obtain the plants that produced spices, the most valuable commodity in the Mediterranean world.

Having left the Canary Islands, he crossed the Atlantic with his small fleet of three tiny ships in thirty-three days, and landed on one of the islands in the Bahamas. For the next three months, the explorers sailed through the islands, first south to Cuba and then eastwards along the coast of Haiti. They encountered all kinds of marvels: dogs that were unable to bark, a 'hideous reptile', which was almost certainly an inoffensive West Indian iguana, which Columbus promptly slaughtered, and men who inserted burning rolled-up leaves into their nostrils and inhaled the smoke.

He had no doubt that he had reached the Indies. There were flocks of green parrots as might be expected, and the local people did not have the black skins and curly hair of Africans but, like true Indians, were olive-skinned with straight hair. Everywhere he went, he picked twigs and leaves from trees and bushes to see whether or not they were aromatic and might have merit as a spice. He showed samples of cinnamon, pepper and other genuinely oriental spices that he had brought with him to the local people, to see if they could match them; but they could not. He and his crew tried the local food and became the first Europeans to taste cassava and maize. He also searched intensively for gold, but only found it in disappointingly small quantities.

After three months of exploration, he set off back to Europe with samples of his discoveries. The King of Spain received him in Barcelona and there, at the Court, he exhibited the ten half-naked, painted 'Indians' he had brought back with him, as well as samples of gold, amber and cotton. He produced cages containing green and yellow parrots and other birds 'never before seen', and samples of dried leaves that he hoped might find favour as spices. Among the still viable plants he had with him were tubers of one similar to the Mediterranean convolvulus, which he declared were excellent eating and tasted 'rather like sweet chestnuts'. This vegetable was planted in gardens in Barcelona and was probably the first New World plant to take root in the soil of the Old. The people of Haiti, where Columbus had collected it, called the tubers 'comoto'. The Spanish turned that word into 'batata' and in English it became 'potato'. Only a century later did it acquire the adjective 'sweet'. Most of the ships returning from the New World over the following decades brought it back as

A seventeenth-century gardening book
John Parkinson, botanist to Charles I, published in 1629 his *Paradisi in Sole*, a pun on his own name since it can be translated as 'Park-in-the-sun'. It was the most popular gardening book of its time and, although most of it is concerned with flowers, it has a substantial section entitled 'The Kitchen Garden' of which this is one page. It shows, in addition to the caraway, three kinds of potato, the Spanish (or sweet), the Virginian (which is today the one called simply potato) and the Canadian. Of this third kind he writes 'We in England, from some ignorant or idle head, have called them Artichokes of Jerusalem, only because the root, being boyled, is in taste like the bottom of an artichoke head: but they may most fitly be called Potatos of Canada, because their roots are in form, colour and taste, like unto the Potatos of Virginia, but greater, and the French brought them first from Canada into these parts'. His name was scarcely an improvement, since the plant belongs to the sunflower family; but 300 years later, we still cling to our earlier error and persist in calling it the Jerusalem artichoke.

1 *Carum.* Carawayes. 2 *Battatas Hispanorum.* Spanish Potatoes. 3 *Papas seu Battatas Virginianorum.* Virginia Potatoes. 4 *Battatas de Canada.* Potatoes of Canada, or Artichokes of Jerusalem.

part of their stores. It rapidly became very popular, since it acquired the reputation for 'inciting to Venus' and 'procuring bodily lust'. Being a tropical plant, it needed warmth, so it did not spread into the gardens of northern Europe; but it was cultivated in many parts of the Mediterranean, and it still is today.

A whole succession of new plants followed it which were to become established in the gardens of the Mediterranean people and were to provide them with a standard part of their diet. Conspicuous among them were species belonging to a huge plant family, the Solanaceae. One of the characteristics of this family is that they produce, in their roots, leaves or berries, chemicals called alkaloids which may be extremely poisonous and produce strange effects on the human nervous system. A few species are native to Europe, and their properties had been known to apothecaries, herbalists and necromancers since ancient times. One, the mandrake, so surrounded by medieval superstitions, contains a sedative and, in the absence of any anaesthetics, was used to alleviate the appalling pain of amputations and other surgical operations. Another, henbane, soothed toothache. Woody nightshade was prescribed for whooping cough, asthma and rheumatism. The fruit of the thorn apple was made into drinks that soothsayers used to induce hallucinations. The juice of the deadly nightshade is not only extremely poisonous, but, dropped into the eye, causes the pupil to dilate. Women used that to make themselves appear more alluring, in spite of the fact that after doing so they could no longer see straight.

But whereas Europe has only twelve native species of this family, the Americas have well over a thousand, and one of the first to be sent back by the plant-collecting explorers was one which provided the people of the Andes with their staple diet. When it arrived, its tubers were thought to be very similar to those of the sweet potato, and it was therefore given the same name. This was the vegetable we call today, without any adjectival qualification, the potato. It arrived around 1570, probably from Chile, though who first brought it to Europe is not clear. Like so many of its family, its berries are extremely poisonous, but its tubers are packed with starch; it must surely be one of the greatest and most influential gifts to come from the New World to the Old. It produced five times as much food from a given area of soil as any other plant known at the time. Some social historians even regard it as one of the main causes of the sudden increase in the size of the population of Europe that occurred at around this time.

The Spanish, doggedly continuing their search for spices, found another member of the family that had brilliant red fruits which, when bitten, produced a painful burning in the mouth. Such a plant, they argued, could quite reasonably be counted as a spice. The one it most closely resembled, they thought, was pepper, which is the fruit

Besler's Pepper
The Prince Bishop of Eichstätt, near Nuremberg in Germany, who ruled at the beginning of the seventeenth century, was an enthusiastic plant collector and owned a famous garden. He commissioned Basil Besler, an apothecary who was also director of the garden, to draw and publish every plant species growing there. This is his portrait of the chilli pepper, which he still describes as coming from the 'Indies'.

of an Indian jungle vine. In truth, there is little similarity between them, but 'pepper' it became and remained. This New World version ultimately proved very popular in its own right and has now been bred into many varieties. Some produce large fleshy fruit, green turning to scarlet, that can be eaten as vegetables; others, small and fiery, are known as chillis; and yet others are dried and ground to produce the pepper known as paprika.

In Mexico, the Spanish acquired another solanaceous species which the Aztecs grew for the sake of its tasty yellow berries and called *tomatl*. This too was relished in Europe and cultivated so that its fruit became bigger and redder. The Italians called it *pomo d'oro*, golden apple, the French *pomme d'amour*, love apple, but ultimately most nations settled for their own approximation of the Aztec name.

It was yet another member of the Solanaceae that produced the aromatic leaves that Columbus had seen the 'Indians' smoking. Tobacco was soon introduced into Spain where at first it was cultivated for medicinal purposes and as an ornamental plant, but before long the extraordinary practice of smoking its leaves, already adopted by the American settlers, became fashionable among the wealthy in Europe. The French ambassador to the Portuguese court in Lisbon, Jean Nicot, purchased some of its seeds and sent them to Paris. There the plant was named Nicotiana in his honour, and when chemists later isolated the alkaloid in its leaves that produced the pleasant addictive effect, that carried his name too.

Besler's Tomato
Besler's catalogue of the garden in his charge was published in 1613 under the title *Hortus Eystettensis*. Its magnificent engravings measure, in the original, 16 inches across. It is interesting that the tomato should be included in it, for the garden and the book were ostensibly concerned not with edible vegetables but with ornamental plants and flowers.

Not all the plant introductions, of course, were members of the Solanaceae. One was related to the broad bean that Europeans had cultivated since prehistoric times. Initially, however, they grew this new bean for the sake of its splendid scarlet flowers. It, too, came from Mexico and was called by the Aztecs *ayacotl*. Eventually people started to eat not only its fat seeds, but the tender young pods, and gave the Aztec name a Gallic culinary sound by turning it into 'haricot'.

A species of sunflower was sent back from Virginia for the sake of its edible tubers. They were considered to have a flavour a little like the artichoke, the large thistle heads that the Arabs had introduced to Spain, so this new plant was also called an artichoke. To distinguish it from the original owner of that name, the Italians called the American plant *girasole*, that is to say a flower which, like its relative the sunflower, 'turns with the sun'. But cooks who knew only the tuber and not the flower soon converted that word into 'Jerusalem', giving the false impression that the plant came from the Holy Land, a misconception that was further reinforced when they called the soup they made with its tubers Palestine Soup.

The Spanish organised the importation of exotic plants in a very thorough way. They established nurseries on the Canary Islands, where the plants were taken as soon as the ships from the New World arrived in harbour and given the special care that many urgently needed after a month-long voyage in temperatures and salty air that did not suit them. Then, when they had recovered and perhaps had been propagated, they were shipped off on the last week of their trans-Atlantic crossing, so that they arrived in Europe in relatively good condition.

Many were imported simply because they were decorative or curious, and few could have seemed stranger to European eyes than the bloated thorn-covered cacti. This family is almost exclusively American and most species live in deserts. Mexico, Arizona and California, then all part of what was called New Spain, possessed them in abundance. Some of these amazing plants arrived in Europe as early as 1565. One of them, Opuntia, flourished particularly well. Its stem is formed by a series of spine-covered oval pads, any one of which will strike roots if it is properly planted. It grew so well and so thickly that it was used to make hedges that were impenetrable to both men and animals. Furthermore, it produced purple fruits that tasted good and were shaped somewhat like pears or figs. In North Africa, it grew so well and spread so fast that some people in Europe believed that their 'prickly pear' originally came from there and that it was more accurately called 'Barbary fig'. Mexico also provided the agave which, like the cactus, is especially adapted to desert conditions and retains its water, not in its stem, but in its rosette of thick fleshy leaves. It, too, was imported at an early date to decorate Mediterranean gardens, but soon it escaped and established itself in the wild. It is sometimes called the century plant in the mistaken belief that only when it is a hundred years old will it sprout the thirty-foot high mast which carries its flowers. In fact it does this after only fifteen years or so; but once it has flowered, the whole plant dies.

Both the agave and the prickly pear are now so widespread around the Sea that many people consider them characteristic plants of the Mediterranean scene. The sad fact is that they are so at home there because they evolved originally to live in deserts.

A third alien plant, the eucalypt, has now joined these two in the typical Mediterranean landscape. It came from an even more distant continent, Australia. There, five hundred different species of them grow in all conditions from humid rain forest to arid desert. One of them, the blue gum, is found in southern Australia in a climate that is not unlike that of the Mediterranean. It is a tall tree, growing up to one hundred and eighty feet, with bark that peels off in strips, giving the trunk a streaked appearance. Like nearly all eucalypts, it grows continuously throughout the year and never drops all its leaves

Mexicans run wild
Agave (*above*) and Opuntia (*below*) are now fully at home in Mediterranean countries, for the hot summers and parched soils closely match the conditions in the semi-deserts of the New World where they evolved.

simultaneously. Furthermore, it grows extremely quickly – up to five feet in its first year and up to thirty-three feet in ten. Its wood is good, strong and heavy, and has considerable commercial value. In 1804, less than twenty years after the establishment of the first British colony in Australia, its seeds were sent to France. There it was discovered that it would grow on deforested land where the soil was so thin and badly eroded that few other trees could find sufficient sustenance. So, throughout the nineteenth century, the blue gum and several other species of eucalypt, many of which hybridise readily with one another, spread on eroding near-desert lands right round the Mediterranean, in North Africa as well as in southern Europe, serving as windbreaks, providing welcome shade and stabilising vanishing soils. Indeed, eucalypts seem to grow even more strongly in their new home than they did in their native land, for the insect pests which attack them in Australia do not occur in the Old World.

Plant parasites and diseases did not, in those days, move easily from one continent to another. The voyage to Australia by sail took three

Flourishing Australians
The several different species of Eucalyptus imported into the Mediterranean area have hybridised, so that precise identification is often difficult. They have become very valuable elements in the countryside, being able to flourish where few other trees can grow at all, and having an ability to survive fire by producing new shoots from resting buds within the bark.

or four months. Even crossing the Atlantic took several weeks, so any plants on a ship that were infected were likely to become visibly diseased and even die before they reached their destination and were, accordingly, probably thrown overboard. The early steamships which were being built at the beginning of the nineteenth century used such vast quantities of fuel that they were unsuitable for long voyages, but by the 1860's their design and efficiency had been improved so greatly that they were introduced on the trans-Atlantic route. The crossing was then reduced to a mere nine or ten days.

In 1863, a new and devastating disease appeared in the vineyards of France. Patches of vines, often in the middle of a field, began to wilt. Their leaves lost their glossy green colour, and turned yellow with a reddish rim around their edges. The fruit, if they bore any at all, remained small and useless for wine-making. Next year, the affected area was greatly enlarged. Those plants that had first developed the sickness lost their leaves and died. When they were dug up, their roots proved to be rotten and black inside. More and more vineyards became infected. No one knew the cause of the disease. Some said it was due to over-pruning, others to soil exhaustion. Whatever it was, no one could find a way of stopping it. As it moved across France, from vineyard to vineyard, real alarm spread among the growers. World-famous vintages, that had been selected, developed and refined over many generations, were threatened with total extinction.

Commissions were set up, first by the wine-growers and eventually by the French Government itself, to establish the nature of the affliction; and huge sums of money were offered as prizes to anyone who could find an effective cure. There were hundreds of suggestions. Some of these were based on little more than medieval superstitions, such as the recommendation that a toad should be buried alive beneath each infected plant. Others were more technologically sophisticated, with prescriptions for a great variety of chemical treatments. All were entirely empirical, since no one was sure of the nature of the disease.

The Government commission, however, began to investigate the problem scientifically. They dug up not only dead vines but also those that were showing the first signs of the disease, and on the roots of these they discovered a bright yellow shiny coating that, on close examination, proved to be a dense mat of tiny aphids known to science as Phylloxera.

The aphid most familiar to us is the green-fly. All members of this vast family of insects feed by means of a hair-thin tube which they thrust into plants. This tube is a very complex structure which enables the insect not only to draw out sap, but also to inject a saliva into the plant which may cause its tissues to grow in a distorted way. Though the insects actively suck, it seems that the sap rises through their mouthparts largely as a consequence of the sap pressure within the plant. When the plant sickens, that pressure falls. The aphids then cannot feed and they move away to another plant. This explains why none of the wine-growers themselves had discovered the insects that were devastating their vineyards. They had only dug up diseased vines when they were sure that the plants were past recovery; and by that time the Phylloxera aphids, being unable to attain any more sap, had moved off to feed elsewhere.

Phylloxera, like many other aphids, changes its shape and character from one generation to another, so that during its complicated life cycle, it may assume a dozen different forms and lay five different kinds of eggs. The species that infects vines is native to the eastern half of North America where it parasitises many of the three dozen or so species of wild vine that grow there. In spring an egg, laid at the end of the previous season on a vine stem, hatches out into a yellow wingless insect that, like all emerging at this stage, is female. She is about half a millimetre long, only a tenth as big as a common garden greenfly. She crawls up onto the upper surface of a leaf, and stabs her feeding tube into it. As she sucks the sap, so she injects a hormone which causes the leaf cells to grow in a different way and produce a dimple in the leaf surface. This develops into a spherical gall that projects below the leaf. The entrance to it is through a small slit in the upper surface, which is protected by palisades of stiff hairs so arranged that it is impossible for any other tiny insect to crawl into the gall, but easy for one to crawl out.

And very many will. The female sits in the hollow gall continuously drinking the plant's sap, moulting as she grows, and laying eggs. Although they have not been fertilised by a male, they nonetheless hatch. Again they are all female. As she continues to lay, so her young clamber out of the gall. In the later stages, it seems that they are in effect pushed out by the sheer quantity of eggs that their mother is producing. The young females crawl away to start feeding for themselves and producing their own galls. Within a few weeks, they too are laying viable eggs, again without the assistance of a male. This process may be repeated many times during the summer. The founding female usually lays between three and four hundred eggs, but as the generations succeed one another, the number of eggs they produce diminishes and the individuals that come from them become slightly smaller in size. Their skin also becomes increasingly roughened by tiny tubercles and their antennae develop deeper notches. By midsummer, some of these insects, now rather different in appearance from the founding female, instead of staying on the leaves, clamber

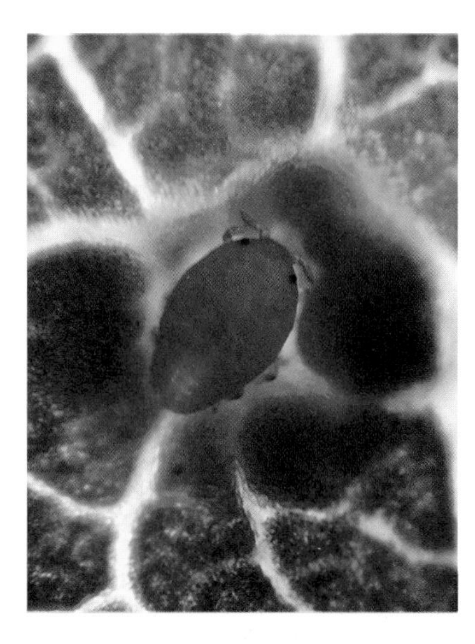

A devastating American
A newly-emerged Phylloxera aphid (*above*) settles down on the leaf of an American vine, surrounded by cells which she will inject with a growth hormone so that they develop abnormally and produce galls. In a heavy infestation, these may cover the entire leaf (*above right*).

down the stem, into the soil and fasten themselves on the vine roots. Once again, they insert their feeding tubes and begin to drink sap, and once again their saliva produces galls. Now begins a whole succession of generations that spread through the roots, just as the previous generations had spread through the leaves.

In early autumn, some of the eggs laid in the root galls produce yet another form of the species – females with wings. These crawl out of the ground and up the stem of their host plant or flutter away to one nearby. There, on the stem, each lays a single egg. The insects that hatch from these eggs include both males and females and, as winter approaches, they mate. The fertilised females each lay a single egg in a crevice in the bark, which will remain there throughout the winter and hatch the following spring.

The vine leaves now die and with them the leaf-living Phylloxera, but the individuals ensconced in the galls on the roots live on through the winter and even survive severe frosts. So many generations have now passed that the single founding female could, theoretically, have given rise to 48,000,000,000 descendants. In practice she will have produced many fewer than that, for all kinds of hazards will have reduced their numbers. Even so, she by herself could have initiated the devastation of a whole vineyard.

This is the life-cycle followed by the insect in North America. In its new European environment, however, it behaved a little differently. For some reason, which even now is not fully understood, the leaves of the European species of vine did not suit it; and, except on very rare occasions, it short-circuited its full American cycle and restricted itself to the root-living phase. The yellowing of the leaves that was the first sign of an infestation was due not to leaf-galls but to the loss of nutriment and water caused by the attacks on the roots. Phylloxera's saliva did not merely induce galls on the roots of the European vines; it killed them.

Nothing of this was known to the French wine-growers. Even after the yellow hordes on the roots had been discovered and recognised, there was argument as to whether the insects were the cause of the disease or whether they were merely taking advantage of plants that had been weakened by some other factor. There were, however, several species of American vine growing in France. One of these importations must have been responsible for introducing the parasite in the first place. They had been imported by plant breeders for use in the development of new hybrids, and none of them produced wine of quality. Now they became of considerable interest for another reason. Although some were infected by Phylloxera, they were seldom killed by it, and several species appeared to be totally immune. One of the scientists investigating the epidemic shrewdly pointed out that Phylloxera could hardly be of European origin for, since it killed the European vine, it would have eventually exterminated itself. It was more likely to have come from America where, over many centuries, the vine species had developed a partial immunity to its injected poisons. By 1868, this identification of the insect and its source had been confirmed. But even though this advance had been made, the problem of dealing with the infection still remained.

One school favoured chemical remedies. Of these, the most effective was carbon bisulphide. Huge engines, belching smoke, trundled through the stricken vineyards, generating the poisonous liquid by passing sulphur vapour over red-hot charcoal and injecting it through nozzles thrust into the ground. Carbon bisulphide is a very dangerous substance. Above a certain temperature, it explodes. Inhaled by human beings in sufficient quantity, it is lethal. In too great a concentration it kills vines. It certainly killed Phylloxera. But its effects were not permanent and the season after treatment, the Phylloxera was quite likely to return.

Another remedy was to flood a vineyard for weeks on end and so drown the insects. This, however, risked damage to the vines and in any case, it could not be used in vineyards on steep slopes, nor in those that were some distance from plentiful water supplies.

Another faction argued for biological control. Since Phylloxera did not parasitise the leaves of the European vines nor the roots of some American ones, why not graft the two together to produce a plant that was totally immune? Nurserymen showed that such grafts would take. Tests were made to demonstrate that the taste of the grapes and

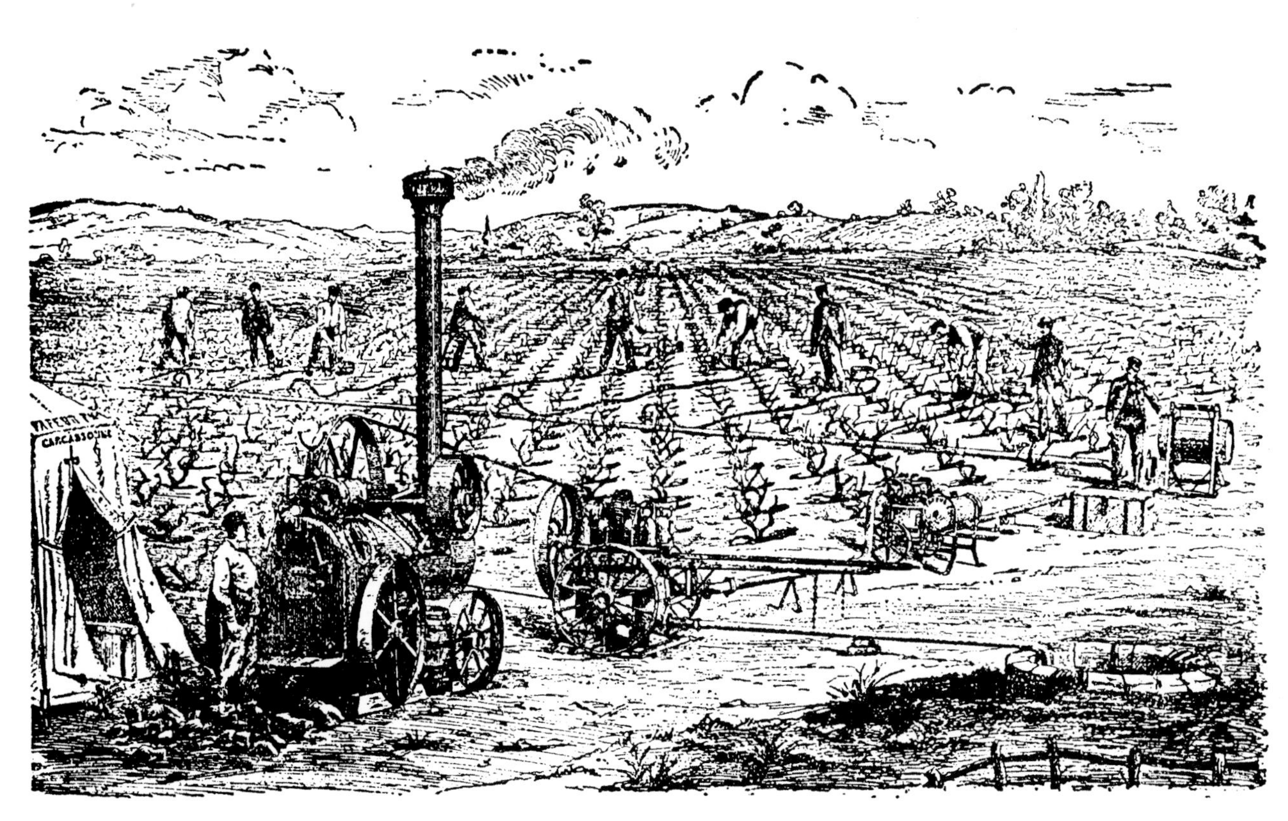

Chemical warfare
In the early stages of the Phylloxera epidemic, the insects that clustered around the vine roots were attacked with a solution of carbon bisulphide, produced by engines that moved through the vineyards drenching the soil. The operators had to take considerable care to prevent themselves being poisoned as well. This illustration comes from a contemporary French newspaper.

the quality of the wine they produced was determined not by the roots but predominantly by the above-ground part of the plant. In 1881, the whole issue was debated at a conference and this solution was accepted by the majority. Not all agreed. Burgundy, jealous of the high reputation of its vintages, passed a law prohibiting the cultivation of any American vines or root-stocks and this remained in force until as late as 1887. But elsewhere, the wine-growers quickly recognised that salvation was at hand.

Over the next few years, more than ten million vine plants were grubbed up from the French vineyards and replaced by as many grafts with American roots. Phylloxera had also by now spread to many other countries around the Mediterranean. It went south to Spain and Algeria, east to Greece and Yugoslavia. But everywhere the technique of grafting stemmed its attacks. The insect that had threatened to deprive Mediterranean man of the drink he had treasured for four thousand years had at last been controlled. But there are still connoisseurs who maintain that nothing today can quite compare with the pure European vintages of pre-Phylloxera times.

THE HUMAN INVASION

Forty years ago, one more animal species invaded the Mediterranean. This time, it came from the north and it was, perhaps, to have the most far-reaching effect of all. The human population of Europe, nourished by the great quantity of crops grown on the lands cleared of forest, had been increasing at enormous speed. In Columbus's time, before the arrival of the potato, it stood at around fifty million. By the end of the eighteenth century it had doubled, and by the nineteenth century it was in excess of four hundred million. The majority of this expanding population did not earn their living in the old way, directly from the land or the sea, but in the industrialised cities of northern Europe. Until the first years of the twentieth century, most of them remained there. Transport was slow, difficult and expensive. Only the wealthy few could afford the time and the money to leave the wetter, cloudier parts of Europe and travel down to the sunny shores of the Mediterranean. A few towns along the coast of Italy and southern France began to build expensive hotels to accommodate them, but their numbers were still relatively few. Then after the Second World War, the introduction of cheap air travel, together with a more equitable distribution of wealth, brought a visit to the Mediterranean to within the reach of the majority of Europeans. And there were now some 490 million of them.

The result was an annual migration of unparalleled proportions. In the summer of 1973, when a detailed survey was made, sixty million people travelled south to the sea. The annual total is now thought to be in excess of one hundred million. Most of them come during the short, three-month summer season and settle within a few hundred yards of the coast.

An inquisitive visitor from another planet might well find it difficult to understand the reasons for this strange mass behaviour. The intentions of the migrants, judged from the literature used by those who persuade them to make the journey, as well as the postcards on which they themselves report on their activities, is to seek empty and lonely villages where classical ruins and Moorish towers dream against a cloudless blue sky. The reality appears very different. Many

Coastline into playground
Cannes (*right*) is near the eastern end of an almost continuous line of buildings that stretches for 200 miles along the French coast. The Camargue (*inset*), at the west end, has no natural beach, and here architects have devised particularly ingenious ways of accommodating the maximum numbers of bodies and boats.

of them spend their time on strips of sand beside the sea where they lie as thickly as seals on breeding beaches or nesting flamingos beside their seaside lagoons.

There they assiduously expose their skins to the sun. The extra-terrestrial observer might find this behaviour particularly baffling. After all, it is a very recently acquired habit. Only a century ago, the wealthy and fashionable ladies who strolled along the Promenade des Anglais in Nice had prided themselves on their milk-white complexions, and had worn clothes of elaborate awkwardness to make it unmistakably clear to everybody that they were totally unacquainted with the outdoor life or with practical matters of any kind. Today, just the same kind of people spend hours each day striving to acquire complexions to suggest that they spend their entire lives out in the open, heavily engaged, to judge from their clothes, in some form of vigorous manual labour.

Sunshine is certainly necessary in small amounts, to enable the body to manufacture Vitamin D, and it also, without any doubt, induces a general feeling of well-being. But its ultra-violet component can be extremely damaging. The human skin has pigment which provides a screen against this danger. It is particularly well-developed among races who come from areas where the sun is very strong, but it hardly exists in the skin of those from cloudier countries. The pigment will of course increase, and so intensify the screen, if the skin is exposed for any length of time to the sun; but whereas Arabs, Asians and Africans develop a dense tan that gives them complete protection, some Celtic people and others from northern Europe may never do so, no matter how much they sunbathe. The damage inflicted on them can be far more severe than a few painful nights and the sloughing of the outermost layer of skin. It can cause skin cancers, and even damage those cells in the skin which resist the development of such growths. Today, in Britain alone, one thousand new cases of skin cancer caused by sunburn are reported every year. A significant proportion of them will ultimately prove lethal.

Nonetheless, each year, the passion to seek the sun attracts more and more people to the Mediterranean. The money to be made from looking after them has induced nearly all the countries around the Sea to modify their economies and revolutionise the lives of their inhabitants. And villages and coastlines that had hardly changed since the time of the Crusaders have been transformed.

The migrating millions
At the beginning of summer each year the beaches all round the Mediterranean suddenly fill with dense assemblies of human visitors. During the course of the season, sixty million visitors will arrive. Four months later, with equal suddenness, the crowds disappear, leaving the beaches deserted and the hotels empty.

The enduring plastic
Unlike organic materials such as wood or paper, the plastic from which so much of our packaging and everyday articles are made today does not decay. So plastic bottles and cartons, sandals and bags, carelessly discarded, remain as almost permanent disfigurements of the beaches.

THE PLUNDER OF THE WILDLIFE

The vast majority of the summer visitors insist on staying close to the sea. Many will complain most bitterly if they cannot see the water directly from the window of their hotel room. Suddenly rocky headlands, barren stretches of garrigue and reed-covered marshes around the mouths of rivers, that no one hitherto had wanted and that had provided quiet refuges for the surviving wildlife, were in great demand. Huge, high-rise hotels were built like gargantuan palisades along the beaches. Swamps were drained to rid them of mosquitos that might bite visitors unacceptably. Stretches of lonely dunes were bulldozed out of the way to create new building sites. Wide motorways were driven along the coast to link one resort with another, and all of them to the airports where giant planes come down to disgorge passengers four hundred at a time.

The visitors themselves, as well as the buildings, roads and marinas needed to accommodate them, inevitably take a toll from the countryside, wearing foot-paths down to the bed-rock, trampling and uprooting wild flowers, carelessly setting fire to the dry summer forests, and littering the countryside with discarded plastic wrapping and other refuse; but their devotion to the beaches reduces their impact on the hinterland to a considerable degree.

Inland the greatest danger to the wildlife comes from the permanent population. They too have greatly increased in numbers in recent years, and many are passionate hunters. They pursue birds of all kinds, large and small, shooting them and trapping them without mercy and often in defiance of the law. The most heartless slaughter of all is that inflicted on the migrants. Every year thrushes and blackcaps, pipits and robins, some near exhaustion after a long journey from Africa and a non-stop flight across the Mediterranean, are assaulted by massed ranks of hunters armed with shotguns and rifles. Captive birds, calling from carefully placed cages, lure them down to traps and snares. Sticks smeared with bird-lime are put in trees so that, when the migrants come down to rest, their feet stick fast and their fluttering wings become fouled.

A few people make a living from doing this. A few more, seeking to justify the massacres, maintain that the tiny bodies make a particularly exquisite pâté. But no one can maintain that the birds are needed for food, or that as they pass through on their journeys, they cause major damage to crops. Men shoot them primarily for pleasure. The slaughter is at its most intense, not in North Africa or the Levant, but in the wealthy countries of south-west Europe: Italy and France, Spain and Portugal. About fifteen per cent of all the migrants are killed. Each year, several hundred million wild birds die at the hand and the whim of man.

In the Sea itself, fishermen harvest other migrants. Each year the

The massacre of the migrants
A nightingale, on its way north from its winter quarters in tropical Africa to its breeding grounds in Europe, arrived in Cyprus and alighted in a tree. But its perch had been smeared with bird lime. With its feet stuck fast, it lost its balance; now it hangs awaiting the return of the trapper who will kill it to make it into pâté.

most magnificent of all the Mediterranean fish, the tunny, swim in from the Atlantic to spawn. They shoal according to age; so all the fish in one group tend to be approximately the same size. Some are immense, around twelve feet long. In places, because of the shape of the coastline and the topography of the sea floor, they have to swim along a relatively restricted and predictable route; and there people wait for them. In Sicily and in Tunisia, the fishermen set long nets hanging from floats and stretching diagonally across the migration path for as much as three miles. When the migrating fish meet the net, they are deflected from their course and swim along the face of it until they enter a long corridor formed by two hanging nets. This leads to a chamber which has not only an end wall but also a netting floor. Once the fish have entered this, the men in boats pull up the free edge of the floor, and the tunny are trapped. As the men pull the floor in, their boats move towards the end of the chamber, shortening it and forcing the fish closer to the surface. Panic then grips the fish and they thrash about desperately, churning the water white. Within a few minutes, they are so exhausted that the men can jump in alongside them, spike them with hand gaffs and heave them into the boats. The biggest of them are far too large for a single man to handle. They are noosed with a rope around their tail and hauled out by winches. A single chamber may trap a hundred or so tunny and provide thirty tons of superb meat; and throughout the two-month season, the trap may be worked once or even twice a day. The slaughter may be horrific, but at least it brings life, both economic and in terms of physical sustenance, to those who wreak it.

Until the last century, fishing techniques in the Mediterranean had changed little since medieval times. Men relied on their intimate knowledge of their own patch of sea and their great understanding of the habits of the creatures that lived in it. They set special traps for octopus, and others for lobster. They sailed out into the coastal waters and, in different places and at different times of the year, they spread nets and lowered hooks on lines.

But when, during this century, motor-driven craft became common, another method became very popular: catching bottom-living fish by dragging across the sea floor a bag-shaped net with its mouth held open by a long heavy beam. At first the Mediterranean trawlers which used such nets kept close to their home ports, like other fishing boats; but, as the demand for fish increased, so they had to go farther out into deeper water to get reasonable catches. To do that they needed bigger boats and larger crews. Still the demand for fish was not met, and before long these new grounds were depleted. So once again bigger vessels had to be introduced to travel to even more distant waters, hitherto largely untapped.

The death of the tuna
A shoal of tuna, migrating into the Mediterranean to spawn, has swum into a huge trap. As the waiting fishermen haul in the netting floor, the great fish thrash about, so exhausting themselves that eventually they can no longer resist the men who jump in beside them and haul them out.

Trawling is at its most intense around Sicily. Small boats still continue to scour the sea-floor within a few miles of the coast; but the catches they bring back are scarcely worth the trouble, for the fish are few in number and pathetically undersized. The bigger ships, some over a hundred feet long, are so large that they can go out for great distances and stay at sea for weeks on end. They are so expensive to run, however, that they are only profitable if, when they return, their holds are full of the most valuable kinds of fish. So when a catch is hauled in, the crew sort through it and throw away all the commoner ones. That can be two-thirds of the entire catch, and, by the time the discarded fish get back to the sea, they are of course dead. So the damage inflicted by these new trawlers is even more widespread than the catches they bring back might suggest; and there is no doubt that if they continue to work in this way, and to increase in numbers as they have been doing, the distant deeper fishing grounds will become as badly over-fished as those inshore.

The harvest of the Sea Fishermen in Morocco bring their catch ashore.

On the opposite African coast, in Tunisia, the situation is very different. There the number of trawlers is still small, and they are forbidden by law to fish in water less than one hundred and fifty feet deep, where many of the bottom-living fish spawn. As a result, the fish they bring back are twice or three times as big as those caught by the Sicilian boats, and ecologists estimate that the weight of living fish, in a given area of Tunisian waters, is twenty times greater than it is around Sicily.

The temptation for Sicilian trawlers to sail right across the Mediterranean and into Tunisian territorial waters is obvious and great. Many do so and are caught by Tunisian gunboats, escorted to ports on the southern shore, and there impounded. Similar trouble is being caused by Italian fishermen in the Adriatic who are sailing into Yugoslavian fishing grounds. Bad conservation is now leading to international quarrels.

The shallow offshore waters of the Sea are vital to the survival of the fish population. The sea-grass, Posidonia, so characteristic of this comparatively tranquil tideless sea that grows thickly here, is an important factor in the health of its waters. Being a green plant, it builds its tissues by harnessing the energy of the sun and in doing so it releases oxygen in extraordinary abundance. A square metre of Posidonia produces ten litres of oxygen a day, which is twice as much as that produced by a square metre of forest. Its leaves, both while they are alive and as they decay, are excellent food for many marine creatures, and its thickets provide invaluable shelter for tiny, newly-hatched fish. Trawling rips up and destroys these beds; so it is prohibited in the shallow waters where the Posidonia grows. Nonetheless, more and more of these meadows are destroyed each year by illegal fishing. They are also being killed in other ways. All around the coasts, silt of one kind or another is being washed down into the Sea. It comes from land newly-stripped of its cover of vegetation, from sites where new hotels are being built and new jetties being constructed to create holiday beaches, and from effluents and sewage that is still being poured, untreated, into the Sea. The silt settles in a thick blanket on the Posidonia. Deprived of light, the leaves die and the meadows rot. So the young fish fry are deprived of both shelter and food, the waters lose much of their oxygen, and fish stocks fall still further.

At the eastern end of the Sea, the fishing fleets of Egypt have also suffered a disastrous fall in their catch. Tragically, the damage here was caused by an ambitious scheme that was intended to increase fertility and bring great economic benefit to the country. In the 1950s, a proposal was launched to build a High Dam across the Nile at Aswan. Water rushing through its turbines would create vast

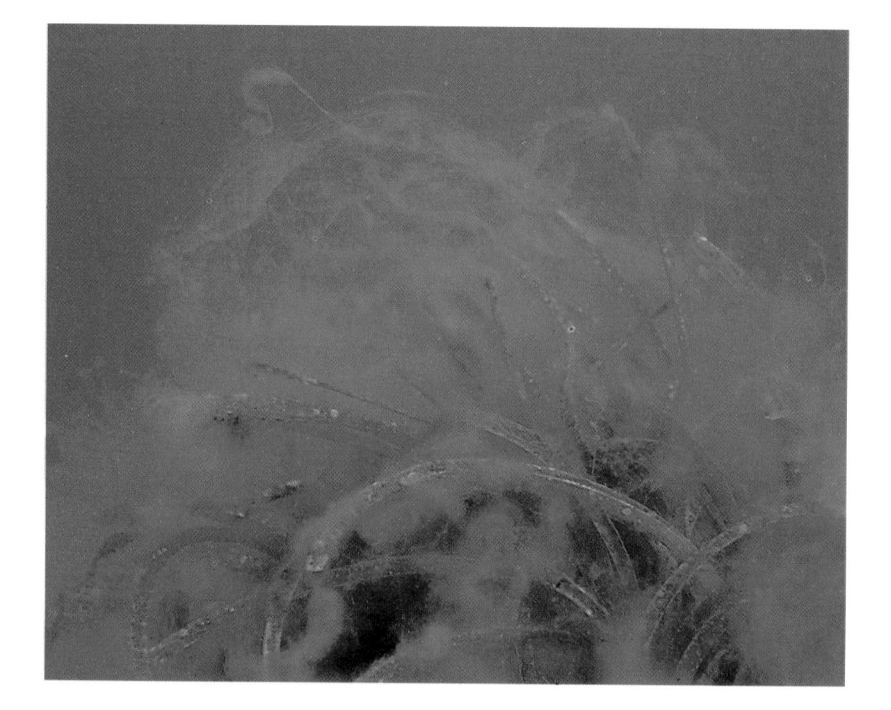

Lethal mud
Like all green plants, Posidonia needs sunlight to live. Screened from it by a blanket of sediment washed down from the land, it dies. Filamentous algae then grow on its dead leaves.

resources of electric power. It would also provide control of the flooding of the river which, in some seasons, had been catastrophic. So the dam was built. It created a gigantic lake that flooded the valley for three hundred miles upstream and a hundred thousand people had to abandon their fields and find somewhere else to live and farm. It did, however, produce the promised electric power and now supplies the country with about half of its needs.

But as the swiftly flowing Upper Nile enters the comparatively still waters of the lake, it drops its sediments, rich in organic nutriments that, for five and a half million years previously it had deposited annually on the lands of Lower Egypt. Furthermore, as its water lies in the lake beneath the fierce sun, twenty-nine per cent of it evaporates. So today, when the engineers release the water through the turbines and it overspills to flood the fields of the lower valley, it no longer spreads as widely as it did during the time of the Pharaohs. Nor does it bring with it the same quantity of nutriment. Artificial fertiliser is now needed in great quantity to replace this and a significant proportion of the electricity generated by the dam is used to synthesise it.

With so little sediment reaching the Lower Nile, the river's delta that extends into the Mediterranean and constitutes Egypt's most valuable land, has stopped growing. In places, the waves of the Sea are actually eroding it away. Far from creating new land for farming

Life-giving mud
Mud deposited from the flooding Nile (*top*) once fertilised the fields of Egypt every year. Now it no longer reaches them, but remains trapped behind the Aswan High Dam (*above*).

as some believed it would, the High Dam has actually reduced it. One authority estimated in 1985 that, taking into account erosion of the delta, reduced flooding by the river, and the loss of fields beneath the lake, the arable land of Egypt had been reduced by 88,000,000 square metres.

The lack of nutritious sediment in today's Nile has one final effect. In the past the Nile waters produced a great proliferation of marine algae in the sea around the delta. That in turn sustained vast shoals of sardines. Now there are no algae and no sardines. One of Egypt's most valuable fisheries has collapsed.

In 1972, the ecological health of the Mediterranean basin seemed to have reached crisis point. The catches from fishing were shrinking. The beaches that brought such wealth from tourists were polluted with refuse and raw sewage. The lands were becoming more and more barren. It looked as though civilised man had, after a mere three thousand years, finally fouled his own nest. Split by ancient religious and philosophical antipathies, divided by economic circumstances, with some of the richest countries facing on the opposite side of the Sea some of the poorest, it seemed that the chances of a united attitude and agreed policies were so slim as to be almost non-existent. But without such an agreement, the accelerating processes of pollution and destruction would continue, and all would be lost.

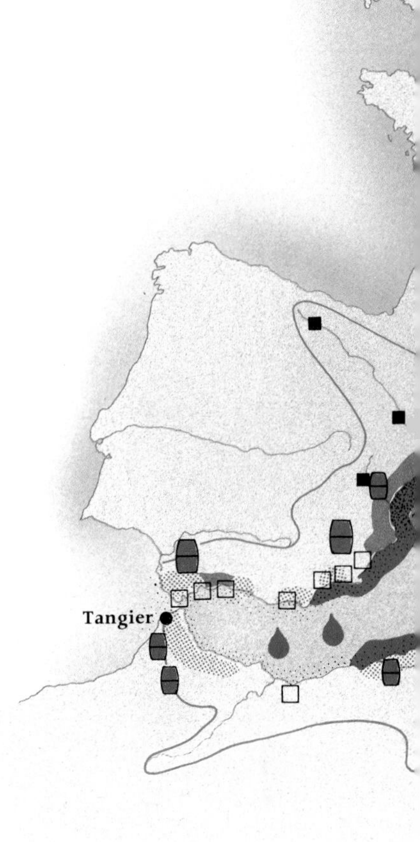

In that year, the United Nations called a conference in Stockholm on the environmental crisis that faced the world. The situation in the Mediterranean was much in people's minds, but they were also deeply troubled by ecological disasters elsewhere in the world. In northern Europe, oil tankers had been wrecked, and crude oil spilled over great areas of sea and long stretches of coastline. In Japan forty-one people had died and seventy-one others had become seriously ill as a result of eating fish caught in Minamata Bay. The source of this poison was found to be mercury that industry had been pouring unchecked into the waters for over thirty years.

As a result of the decisions taken by this conference, a new agency was set up, the United Nations Environment Programme. One of its first acts was to call a new meeting to deal with the crisis in the Mediterranean. This duly took place in Barcelona in 1973. Sixteen out of the seventeen states with Mediterranean coastlines attended. Only Albania was missing.

In spite of a great deal of evidence, the participating nations were not convinced that it would be economically profitable for them to take action there and then. So instead, they agreed only to seek more detailed scientific observations about what was happening in the Sea and to agree on the answers to such apparently simple, but in practice difficult, questions as how pollution should be defined, how it should be measured, where it came from, and how it circulated. Unless such facts were discovered, and universally accepted, the framing of international agreements would be impossible. It took a long time. Two hundred research stations were set up all around the coast to measure pollution and monitor its changes, whether for better or for worse. Their findings were horrifying. Over ninety tons of pesticides, which are remarkably stable and can accumulate in the bodies of animals until they reach lethal levels, were washed into the Sea from the land. Eight hundred thousand tons of oil were dumped in it, partly from accidental spillage, and partly from the holds of tankers that were being washed out at sea. Over four hundred and thirty billion tons of domestic waste and sewage were flushed into it, ninety per cent of which was untreated. Coastal industries contributed over five thousand tons of zinc, over one thousand four hundred tons of lead and over ten tons of mercury.

Painstakingly, blame was apportioned and guilt acknowledged. Complex bargains and trade-offs were negotiated. Technologically advanced countries on the northern shore, with their industries in full production, pleaded that the proposed strict standards of pollution control should apply only to those installations that were about to be built, rather than those already in operation, which would be ruinously expensive to modify. The developing countries, which had

The polluted sea

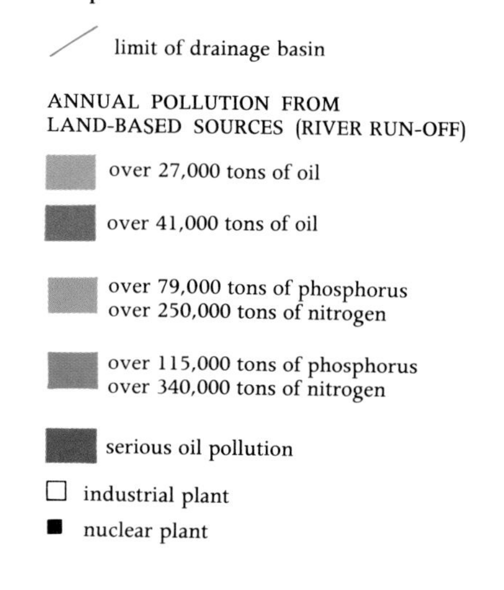

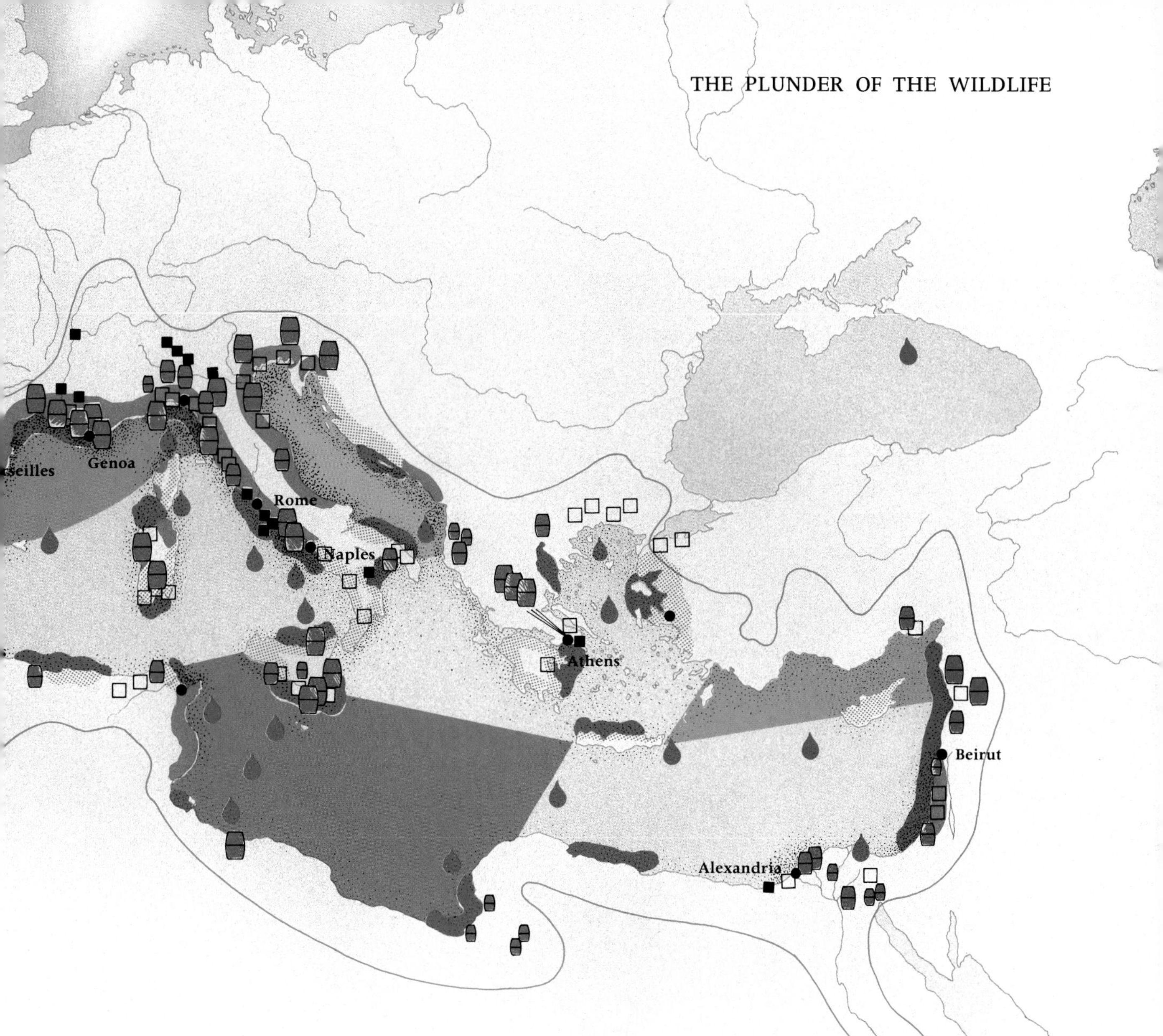

● population centres over 750,000

Tourism

moderate density

high density

Sewage and waste output

low

medium

high

Oil refineries

over 100,000 barrels per day

20,000–100,000 barrels per day

under 20,000 barrels per day

major oil tanker spills (1970–1980)

still to establish their industrial plants, agreed, provided that the legislation was based on the percentage pollution of coastal waters rather than the emission level from the factory. The dumping of oil and any other waste at sea was universally outlawed. A Centre, paid for internationally, was set up in Malta with specialised equipment and trained staff to deal with any oil pollution emergency. Athens, which had one of the most filthy polluted bays in the whole Sea, started to build itself a proper sewage system. The huge petro-chemical and steel-making complex west of Marseilles installed equipment that cut the pollution it created by ninety per cent. Many problems remain to be solved before the Mediterranean can be reckoned to be healthy once again. Fishing is still not properly con-trolled. But a start has been made.

THE RELICS OF EDEN

The lands around the Mediterranean have been changed far more radically by man than has the sea. In classical times, lions lived in Greece, as we know from the writings of Herodotus and Aristotle. The wolf and the bear, once common around both the northern and the southern shores, have now been exterminated over much of their former range. The Spanish ibex, a wild goat with spectacular curved horns, originally lived all over the peninsula. It was hunted with such intensity that eventually its population was split into several small communities, each of which differed slightly from the others in size and coloration. The last of the Portuguese race was shot in 1892, and none has been seen in the Pyrenees since 1907. A few still survive in central Spain. The Spanish lynx, deprived of forests in which to hunt, is now reduced to a group of about thirty individuals in the marshes around the mouth of the Guadalquivir River in Andalucia, and another small population of about the same size in a reserve in central Spain.

Although these are sad losses, other animals closely related to all these species still survive in the wilder parts of adjoining continents. Identifiable remains of the Greek lions have not been discovered, but they were almost certainly very similar to those that are still abundant in Africa. Wolves and bears remain in some numbers in mainland Greece and Yugoslavia. The ibex is now carefully conserved in central Spain and in the Alps; and the northern form of the lynx, which is a little bigger and not quite so handsomely spotted as the Spanish one, is still a common animal in Scandinavia.

The Mediterranean species that are most at risk are its endemics, those that live nowhere else. If they disappear from their present home, they are lost for ever. There is one such endangered endemic mammal, the monk seal. Just how many individuals of the species there were two thousand years ago is impossible to tell, but they were abundant enough for a Greek city on the coast of Turkey to call itself Phocaea – Phoca being Greek for a seal – and to use the animal as an emblem on its coinage. They were still common in Columbus's time. But as the human population grew and fishing became more intensive, men decided that the seals could no longer be allowed to take a share of the fish, and they killed them whenever they had the chance. When, as a result of over-fishing, catches became smaller and smaller, the fishermen put the blame increasingly on the seals and felt even more aggrieved when one got entangled in their nets and caused expensive damage. Furthermore, seal meat and seal skin both fetched good prices, so seal hunting continued with increasing intensity.

Other pressures drove the monk seals from their breeding grounds. Like all other seals, they have to come ashore to give birth to their pups. They need a sandy beach at the head of a gently sloping seafloor if their helpless new-born young are not to be carried away by

Europe's most endangered carnivore
The lynx that lives in Spain, sometimes called the pardel, is slightly smaller and has much more distinct and widespread spots than those that live in Scandinavia and other parts of northern Europe. Once it was widely distributed in the south, but it was exterminated in Italy by 1910 and the last French animal was shot in the Pyrenees in 1917. It is still found in some numbers in the mountains of mainland Greece. Two populations survive in Spain, one in the Coto Doñana and the other in the mountains in the centre of the country.

waves in rough weather and drown. But as visitors to the Mediterranean increased, fewer and fewer such places remained empty and quiet enough to suit the seals. So they were forced to change their habits. Now they pup in dark sea caves that can only be entered by a small boat. Probably less than three hundred and fifty of them survive.

The birds of the Mediterranean have suffered a similar fate. In Roman times, the Dalmatian pelican was widespread and lived as far north as Germany and the Netherlands. Today it is found only in Yugoslavia, Greece and Turkey. The sacred ibis, once so revered and so assiduously mummified by the ancient Egyptians, is now never seen in the Nile valley, and only survives in any numbers south of the Sahara. Its relative, the bald ibis, a bizarre-looking bird with a naked red head and a fringe of long plumes around the back of its skull, is in even greater danger, for it has no tropical headquarters to which to

Two endangered species
Two kinds of pelican nest around the shores of the eastern Mediterranean. The Dalmatian (*above*) is much rarer than the White and is distinguished from it by a lack of prominent black patches on its underwings. The ibex (*right*), a kind of wild goat, was once much hunted for the sake of its magnificent horns. Now it is extremely rare, except in those reserves where it is carefully protected.

retreat. It favours drier country than most of its relatives, and nests communally on cliffs. Once it did so in Germany and Austria, Syria and Algeria. Now eight birds linger outside a small village in Turkey. The only others are scattered in remote parts of the cliffs of Morocco. Its decline is due to insecticide poisoning in the 1950s, and to disturbance of its nesting sites by humans.

Three bird species are endemic to the area: a unique warbler in Cyprus, a nuthatch in Corsica, and a relation of the herring gull known as Audouin's gull that has green legs and a red bill with a black-and-yellow tip, and which nests in one or two places on the mainland and several of the islands. A fourth bird, Eleonora's falcon, migrates in winter to East Africa and Madagascar, but it breeds only in Mediterranean countries; the sight of these superb birds swooping and shrieking around their nesting cliffs is one of the great ornithological rewards of the Greek islands. The populations of all these species are very small and so always vulnerable, but they do not at the moment appear to be in danger.

Distinctive bills
Audouin's gull (*above*) which, apart from colonies along the Atlantic coast of Morocco, is exclusively a Mediterranean bird, is closely related to the herring gull and differs from it by the striking coloration of its beak. The bald ibis (*right*) is the only member of its family to have a bright red beak. It is now so rare that its continued survival must be in doubt. These are members of the last colony in Europe, in Turkey.

Endemic plants exist in great numbers. One of the rarest is known as the Maltese fungus. It is not, in fact, a fungus at all but a parasitic flowering plant that for most of the year lives invisibly below ground, attached to the roots of tamarisk and of one or two other plant species that habitually grow on sea cliffs. In the summer, strange red spikes several inches high, unaccompanied by any leaves, protrude through the earth. They are made up of several hundred tightly-packed flowers. As they seed and die, the spike dries and turns black. The misguided logic of medieval apothecaries led them to suppose that since this extraordinary apparition was red, it must be an excellent medicine for diseases of the blood, and haemorrhages of one kind or another, and, in view of its phallic shape, for genital disorders as well. The fact that it was very rare only made it the more precious. One colony was discovered by the Maltese Knights of St John, living on the top of an isolated sea stack on the coast of Gozo. A permanent guard was put on it. The rocky sides of the stack were smoothed to make the climb up it more difficult, and anyone who was caught stealing a

A rare apparition
The Maltese fungus is only visible above the ground for the few weeks in the year when its flowering spikes appear. For the rest of the time it exists entirely below ground, drawing its sustenance parasitically from the roots of other cliff-dwelling plants.

plant was sentenced to serve in the galleys. The flower-spikes were sent as special gifts by the Grand Master of the Knights to European kings. The plant still grows today on the top of Fungus Rock, and there are other small colonies in Majorca and the south of France.

The Maltese fungus, judging from its present distribution, must once have been widespread around the Sea; but most endemic plants in the area have come into existence because of their isolation on islands. The characters that differentiate them from mainland forms, though distinctive, are often small and technical, so that identifying them may be a job for an expert. Majorca has between forty and fifty such endemic species, Cyprus about seventy, and Crete about one hundred and thirty, which is about one in ten of all its native plants.

The most spectacular animal endemics evolving on islands must certainly have been those now extinct species, the pygmy elephants, tiny hippopotamuses, and the other bizarre creatures that evolved on Malta, Majorca and the other islands five million years ago. Excavations have now shown that Myotragus, the antelope-turned-rodent that lived on Majorca, survived well after the arrival of human beings on the island around six thousand years ago; it may indeed be that this event was ultimately the cause of its disappearance. But one species of animal that probably evolved around the same time is still alive.

In 1977, the sub-fossil bones of a hitherto unknown frog or toad were discovered on the island. Three years later, some strange tadpoles were spotted in a pool in a mountain torrent. The next year, after a careful search, the adults were discovered. Their bones matched those of the fossils and they proved to be a kind of midwife toad. The European species, from which the Majorcan animal is derived, gets its name from the fact that the male takes the fertilised eggs from the female and carries them around entangled in his legs until such time as they are ready to hatch. Then he goes down to water and allows the young tadpoles to swim away. The Majorcan midwife looks after its eggs in the same way, but differs in several other characters. Not only is it slightly different in colour and shape, but it lacks glands in the skin which the mainland toad possesses and which produce a poisonous secretion if the animal is interfered with. It also lays only eight to twelve eggs, which is about a quarter of the number produced by the European midwife. Both these characteristics seem to be the result of living in an environment where it has no predators.

Since the Majorcan midwife toad evolved, however, the viperine snake, a close relative of the grass-snake and very similar to it in appearance, has been introduced on the island. Its main food is amphibians: the Majorcan midwife had no poison with which to defend itself, nor did it produce sufficient young to compensate for this sudden predation. So it was exterminated from most of the island.

The survivors live high in the mountains in permanent pools at the bottom of cascades tumbling through steep-sided gorges over vertical rock faces. It seems that no snakes have so far managed to reach these refuges. How many toads there are today is not certain, but there are unlikely to be, all told, more than five hundred pairs.

Rarities, however, whether they are endemic island species or relicts of populations that once ranged over the entire region, are perhaps of interest largely to specialists. The sights and experiences of the natural world that move us all, whether expert or inexpert, come from whole ecological communities, patches of the countryside where the hand of man has hardly had an effect – forests complete with all the mammals and birds, insects and wild flowers that belong to them, marshes thronged with vast assemblages of waterfowl, rivers rich in fish with beavers and kingfishers and dragonflies. A few such places still exist, largely because until now man had little practical use for them. They were too remote or too difficult to penetrate to be bothered with, as long as easier country was still unexploited. But that is no longer the case. Shortage of land of any kind, and increasingly powerful machines for shifting earth, felling trees and draining marshes, have changed the situation radically.

The Coto Doñana in Spain is one of the richest of these surviving wild areas. The sea that laps its shores is not, it has to be said, the Mediterranean but the Atlantic, for it lies around the mouth of the Guadalquivir River that reaches the coast just west of Gibraltar; but climatically and biologically, it is wholly Mediterranean in character. There live the last Spanish lynx. Skinks and burrowing lizards skitter across its sand dunes. Imperial eagles hunt rabbits, and black vultures search for carrion. Azure-winged magpies and hoopoes nest in its woods. There nearly half of all European species of birds can be seen. Another magnificent marshland still survives around the mouth of the Rhône. This is the Camargue, a wilderness of marshes and shallow lagoons, of dunes and stone pines, the home of the most northerly regular colony of flamingos, where bulls and horses are once again running wild. On the African shore, a third wetland, around Lake Ichkeul in Tunisia, provides winter quarters for two hundred thousand ducks and all the European population of genuinely wild greylag geese. Birds come here from the Netherlands, Finland, Czechoslovakia and Kazakhstan.

The Majorcan midwife
This rare toad, unique to Majorca, was once widespread on the island. Snakes, introduced centuries ago from the mainland, exterminated it over much of its range, and now it survives only in a few isolated pools of just five torrents in the high mountains. If a single snake succeeded in climbing up the rocky gorge to one of these pools, it could exterminate a colony almost overnight.

Virtually all the ancient deciduous forests of the Mediterranean have now been felled, for the soil they generated around their roots was too rich and fertile for it to be left uncultivated. Yugoslavia still retains a few patches of virgin forest. In Plitvice Lakes National Park, ground orchids, lilies and peonies grow beneath beech, fir and spruce. Bears and wild boars wander through them and otters live in the rivers. In Majorca, where ten million people spend their holidays each year, black vultures and osprey nest in the pines growing on remote and inaccessible slopes, and on cliffs facing the sea. In Morocco's Atlas mountains, macaque monkeys clamber among the cedars and scamper across the ground in search of roots.

The Northern Sporades in the Aegean are among the most unspoiled of all Mediterranean islands. Pine forests still cover parts of

them, and the seas around their coasts are still rich not only in fish, but also in many other marine organisms that continuous fishing over centuries have eliminated elsewhere. The remotest of them, Piperi, twenty five miles from the nearest inhabited island and sixty miles from the mainland, is inhabited by 350 pairs of Eleonora's falcons, probably half the Greek population, which wheel and swoop continuously all around its cliffs. The Mediterranean's own endemic gull, Audouin's, has a colony there. Most important of all from a conservation point of view, this is the last refuge of Europe's rarest mammal, the Mediterranean monk seal. We still know comparatively little about this species, so rare has it now become, but it seems certain that two or three adult females live permanently in the waters around the island. They, perhaps with several others that come from farther afield, visit the low, dark shingle-floored caves that the sea has carved in the bottom of the sheer limestone cliffs to give birth to their single pups.

Two Mediterranean endemics
Eleonora's falcon (*left*) is named after a 14th-century Sardinian princess who introduced a law protecting nesting hawks and falcons. It lives in colonies on rocky cliffs and, apart from outposts in the Canaries and north-west Morocco, is confined to the Mediterranean. So too is the monk seal (*right*), which has been so reduced in numbers that it is now Europe's rarest mammal.

THE COTO DOÑANA

This stretch of sandhills, pine forest and marsh extends for twenty-five square miles across the delta of the Guadalquivir River in southern Spain. For 350 years, it was the hunting reserve of the Dukes of Medina Sidonia, and so remained isolated and totally wild, accessible only on horseback. Today it is increasingly threatened by holiday developments around its boundaries, but it still provides homes for some 200 different species of birds, including hoopoe (*1*), Spanish imperial eagle (*2*), spoonbill (*3*), azure-winged magpie (*4*), and squacco heron (*5*).

1

2

3

4

5

THE CAMARGUE

The Camargue is a vast area of marshes, freshwater lakes and saltwater lagoons created by the muds and sands brought down by the River Rhône from central Europe, and deposited here on the coast as the river enters the Mediterranean. Although there is a great deal of cultivation of rice and other crops, and the area is criss-crossed by roads, and in spite of the great numbers of people who come here during the season to hunt the vast flocks of waterfowl, much of the Camargue is still very rich in wildlife. In the marshes, herds of black cattle (*1*) mix with white horses (*4*), both of them descendants of domesticated stock which, having now become totally wild, give a hint of the ancient herds that once roamed over all Europe during prehistory. Both animals live an almost amphibious life, splashing through the shallow lakes and even grazing on aquatic plants beneath the surface of the water. Ant-lions (*2*) and tree frogs (*3*) haunt the reed-beds. As well as the famous colonies of flamingoes that nest in the lagoons (*page 37*), many other birds breed here, including (*overleaf*) bee-eaters (*5*), black-eared wheatear (*6*) and the collared pratincole (*7*) which nests nowhere else in France.

2

3

1

4

5

6

7

PLITVICE

The mountains along the Adriatic coast of Yugoslavia are built of limestone. Rainwater falling on them dissolves away the rock and so disappears into long underground galleries and caverns. Eventually it emerges again in clear rivers laden with lime in solution. This it redeposits as beds of travertine, which eventually build up into dams, over which the waters of the river cascade. In Plitvice National Park (*1*) this process has created a series of spectacular waterfalls linking a chain of sixteen lakes. The surrounding land is covered by one of the last stretches of truly virgin forest to survive on the European shores of the Mediterranean. Beneath stands of beech, maple and fir grow a great variety of lime-loving flowers such as lizard orchid, (*2*), Martagon lily (*3*) and false helleborine (*4*).

1

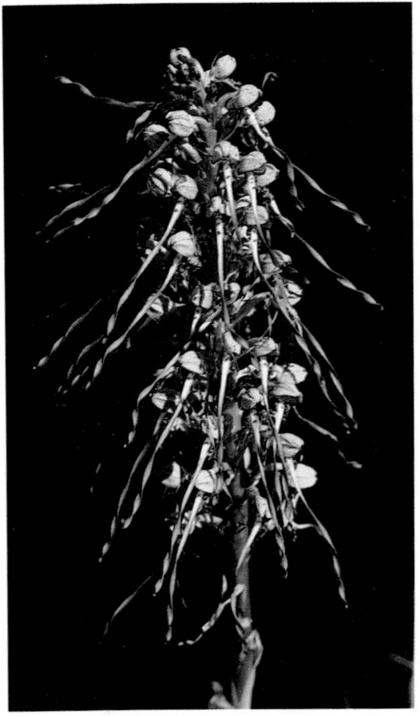

2

3

4

Plitvice lies fifty miles from the Adriatic coast and was declared a national park by the Yugoslav Government in 1949. It is rich in animals of all kinds. Those magnificent birds, the eagle owl and black woodpecker, live in its forests. Wild boar (*1*), which are still to be found in many parts of Europe and are intensively hunted there, are protected and relatively common in Plitvice. They particularly relish the wallows beside the lakes. There are also wolves, deer and a few lynx. The brown bear (*2*) the largest of all Europe's land mammals, is also here. Apart from those in northern Scandinavia, only four isolated groups of these impressive animals still survive in western Europe – in the Cantabrian mountains of north-western Spain, in the Pyrenees, the Italian Alps around Trento and the Abruzzi in central Italy. In none of these places are there more than two hundred or so of these animals, and in the Alps probably no more than a dozen still survive. But in Plitvice they continue to flourish in some numbers, and can frequently be seen bathing in the lakes.

1

2

MAJORCA

Although so many visitors come every year to the small island of Majorca, it still has large areas where wildlife flourishes. The black-winged stilt (*1*), with its spectacularly long legs, which is a rare visitor to Britain, breeds on its marshes. In the mountains of the north, the black vulture nests (*2*). This bird has been steadily decreasing in numbers over the past century, because of persecution and because the carrion on which it feeds is no longer commonly found in the countryside.

1

2

NESTOS

The River Nestos runs south from the mainland of Greece into the Aegean. The constant supply of fresh water running through the channels of its delta, together with the heat of the Mediterranean summer, produced a thick forest that was almost tropical in character. It was in places such as this that the wild grape vine originally grew. Much of the forest has now been felled and the land used for intensive chemical-fed agriculture; but, in the fragments that remain, the wild grape vine still clambers over the trees, together with other climbing plants such as clematis and wild hops. Orchids, lilies and yellow flag (*2*) grow beneath; and water birds, such as the little bittern (*1*) nest in the reed beds beside the water.

1

2

So the Mediterranean still has some remnants of its former glories. In spite of everything, no Mediterranean bird has yet become totally extinct. Only one species of mammal, the wild bull, has disappeared, and even that can be reckoned to survive in the form of domesticated cattle and in the herds that have returned to the wild in the Camargue. But if these wild areas where they live are to survive for much longer, we can no longer rely on the accidents of geography and the oversights of developers that have so far been their salvation. We now have to protect them actively. Many sites are now reserves and national parks; but some of them were given this status at a time when the hunger for land was not as intense as it is now, and the new demands are straining their legal defences to breaking point. Their guardians will need great determination to maintain them.

The Mediterranean is, arguably, the oldest humanised landscape in the world. This is where we first learned to tame animals and cultivate plants. This is the place where we became so powerful that we were able to transform landscapes wholesale, where we started to keep exotic animals and plants brought from all over the world, where we turned half a continent into a market garden.

Might this also be the place where we really begin to learn from our mistakes? As we holiday in its ruined lands among temples that were overwhelmed by desert sands two thousand years ago, and walk across beaches fouled with oil and dying fish that we polluted only yesterday, perhaps we may perceive how much we have lost and come to our senses. Here, certainly, the many nations that share the Sea have recognised that unless they collaborate and agree on how to care for it, everyone will suffer. Even in the Middle East though we are still fighting with bombs and bullets the battles that started in the Middle Ages with bows and arrows, people from both sides are now working together to save the most precious of our possessions, the natural world.

Perhaps that lesson will be the last great gift that the people of the Mediterranean have to offer. One thing is certain. The processes that started here ten thousand years ago and brought the Mediterranean to its present condition are now at work all over the earth. Now it is not just a small sea, and the lands that surround it, that the nations must come together to save. It is the planet.

* * *

ACKNOWLEDGEMENTS

The four parts of this book deal in turn with, broadly speaking, natural history, archaeology, history and ecology. Few people can claim to be equally knowledgeable in all four disciplines. Certainly I do not. I have therefore relied heavily on written sources in order to try and disentangle the story of humanity's attitude to, and relationship with, the natural world. Those works which I have consulted to a considerable degree I have listed in the bibliography.

I have also been greatly helped by several scholars who have been very generous with their knowledge. Dr Graham Drucker introduced us to the troop of Barbary macaques that he had spent several years studying in Morocco. Professor Kenneth Hsü, who was primarily responsible for the discovery that the Mediterranean had once evaporated, and whose book is a vivid and exciting account of the work that led to that discovery, was particularly helpful, as was Professor Harry Smith who guided us through the intricacies of Egyptian history and theology.

During our travels around the Mediterranean, many people solved difficulties that we might, by ourselves, have found insuperable and enabled us to see things we would otherwise have missed. They include: in Cyprus, Andreas Demetropoulos; in Egypt, Dr Ibrahim Muhammed; in Greece, Kostas Christou and Professor Nikos Margaris; in Majorca, Dr Joan Mayol; in Malta, Joseph Vella-Gaffiero and Anthony Parnis; and in Spain, Dr Ramon Viñas.

This book was written at the same time as four films for BBC Television were being made on the same subjects, and my biggest debt of all is to the production team with whom I worked on the project. Their names are listed in full on the right but I owe special thanks to those who researched the subjects we filmed (and many that in the end, we did not) – Neil Nightingale who dealt with the biological and ecological basis of our story, and Karen Bass and Adaire Osbaldeston who investigated the archaeological and historical aspects of it. None of us, however, would have managed to reach our various destinations without the shepherding of Diana Richards, and we were all guided and urged on by the producer and shaper of the entire series, Andrew Neal.

Converting our narrative into print while simultaneously filming it – and furthermore, doing so in time for this book to appear as soon as the films were ready for transmission on television – demanded a degree of hard labour, expertise and patience far beyond that normally expected from a publishing house. Jennifer Fry found the illustrations, Bridget Morley designed the pages, Elizabeth Winder arranged for them to be set and printed, and Crispin Fisher took charge of the whole process from beginning to end. So, miraculously, it was done.

I am very grateful indeed to each and every one.

TELEVISION
PRODUCTION TEAM

Producer
Andrew Neal

Assistant Producers
Neil Nightingale
Adaire Osbaldeston

Research
Karen Bass
Nigel Marven

Production
Diana Richards
Jill Hirschmann

Cameramen
Martin Saunders

Rodger Jackman
Alan McGregor

Huw Davies
Jeremy Humphries

Film Editing
Martin Elsbury
Peter Brownlee

Andrew Naylor
David Thrasher

Sound Recordists
Lyndon Bird
Keith Rodgerson
Graham Ross

PICTURE CREDITS

PHOTOGRAPHS: Page 2 Ardea (J.-P. Ferrero); 8 Robert Harding; 11 David Attenborough; 12–13 Explorer (C. Delu); 14 Bru Coleman (Fritz Prenzel); 14–15 A. C. Waltham; 16–17 National Remote Sensing Centre, RAE; 18 NASA; 20–21 *bottom* OSF (Peter Parks); 21 *right* Biofotos (Heather Angel); 22–23 Ardea (P. Morris); 24 *top* Bruce Coleman (Jane Burton), *bottom* Planet Earth (Christian Petron); 25 Ardea (R. & V. Taylor); 26–27 *top* Planet Earth (James King); 26 *bottom* OSF (Len Zell); 27 *bottom* Bruce Coleman (WWF/Trotignan); 32 ZEFA; 35 Ardea (G. K. Brown); 36 *left* Bruce Coleman (M. P. Price), *right* NHPA (M. Danegger); 37 *top* Bruce Coleman (G. Ziesler), *bottom* Robert Harding; 38 all Smith/Polunin Collection, except *bottom right* Nature Photographers (Paul Sterry); 39 *bottom* Bruce Coleman (John Markham); 40 *top left* Ardea, *top right*, *bottom left* Bruce Coleman (Hans Reinhard), *bottom centre* S. P. Nicholls, *bottom right* Anthony Huxley; 41 S. P. Nicholls, except *bottom right* Bruce Coleman (R. K. Murton); 42 *left* Bruce Coleman (Eric Crichton), *right* Jennifer Fry; 43 *top* Biofotos (Heather Angel), *bottom* Bruce Coleman (M. P. Price); 45 *left* NHPA (Stephen Dalton), *right* Planet Earth (J. & G. Lythgoe); 46, 47 Andrew Neal; 49 *top* Nature Photographers (S. C. Bisserot), *centre* Andrew Neal, *bottom left* Bruce Coleman (Hans Reinhard), *bottom right* Natural Science Photos (G. Mattison); 50 *left* NHPA (Stephen Dalton), *right* Ardea (Ian Beames); 51 all Bruce Coleman (*top*, Frieder Sauer, *bottom left* A. J. Mobbs, *bottom right* Udo Hirsch); 53 Andrew Neal; 54 Ardea (J.-P. Ferrero); 55 Ardea (André Fatras); 56 NHPA (Anthony Bannister); 57, 58–59 Andrew Neal; 61, 62 David Attenborough; 64–65 John Hillelson (René Burri); 66–67 John Hillelson (P. Vauthey/ Sigma); 72–73 Andrew Neal; 74, 75 Michael Holford; 76 David Attenborough; 78 Roger Wood; 79 Ardea; 81 Bruce Coleman (Norman Myers); 82 John Hillelson (Brian Brake); 83 Andrew Neal; 84 *bottom left* Colorphoto Hinz, *bottom right* John Hillelson (Brian Brake); 84–85 Bridgeman Art Library; 85 *bottom left* Michael Holford, *bottom right* John Hillelson (Brian Brake); 86 *top* Werner Forman Archive, *bottom left* ZEFA, *bottom centre* and *right* Robert Harding (Walter Rawlings); 87 all Michael Holford, except *bottom left* John Hillelson (Brian Brake); 88 Photoresources; 90–91 Michael Holford; 92–93 ZEFA (K. Scholz); 94 *top left* Bridgeman Art Library, *bottom left* Anthony Huxley, *right* Spectrum; 96 Ashmolean Museum; 97 Andrew Neal; 98, 99 *top* Department of Antiquities, Nicosia; 98–99 *bottom* Photoresources; 100 *top* Spectrum, *bottom* Robert Harding; 102 *top*, *centre* Photoresources, *bottom* Peter Clayton; 102–103 Spectrum; 103 *top* Peter Clayton; 104 Michael Holford; 106 Peter Clayton, 107, 108 Spectrum; 109 David Attenborough; 110–111 Michael Holford; 111 John Hillelson (Erich Lessing); 112, 113 Andrew Neal; 114–115 Roger Wood; 115 John Hillelson (Brian Brake); 116–117 Andrew Neal; 118–119, 120 ZEFA (K. Benser; J. Behnke); 122–123 John Hillelson (Erich Lessing); 126 Bruce Coleman (Leonard Lee Rue III); 128 British Museum (John Webb); 129, 130 Michael Holford; 131 Werner Forman Archive; 132–133 ZEFA; 133 *bottom* Andrew Neal; 135 Bibliothèque Nationale, Paris; 136 ZEFA (B. Kappelmeyer); 137 Spectrum; 138–139 Andrew Neal; 139 Spectrum; 141 Bayerische Staatsbibliothek München; 142–143 NHPA (Stephen Dalton); 145, 146, 147, 149 and 150 *bottom*, all British Library; 150–151 Bibliothèque Nationale, Paris; 153 Michael Holford; 155 *top* ZEFA (Havlicek), *bottom* Andrew Neal; 156 *left* David Attenborough; 156–157 ZEFA (Konrad Helbig); 161 Ardea (Liz & Tony Bomford); 162 Scala; 164 Jon Gardey; 165, 166–167 Andrew Neal; 170–171 Robert Harding; 174 Nature Photographers (Roger Tidman); 176–177 NASA; 180–181 *top* BBC Hulton Picture Library, *bottom* Mary Evans Picture Library; 184, 185 Royal Botanic Gardens, Kew; 187 *top* Nature Photographers (S. C. Bisserot), *bottom* Robert Harding; 188 Ardea (P. Morris); 191 *left* Alastair MacEwen, *right* Ministry of Agriculture, Fisheries & Food (Crown ©); 194–195 Tony Stone Worldwide; 195 *top* Aspect Picture Library (Derek Bayes); 197 *left* Robert Harding, *right* Bruce Coleman; 199 Ardea (Ian Beames); 201 Andrew Neal; 202–203 Nature Photographers (Roger Tidman); 204 Planet Earth (Christian Petron); 205 *top* Robert Harding (Walter Rawlings), *bottom* Aspect Picture Library (Alex Langley); 209 Ardea (J.-P. Ferrero); 210 Bruce Coleman (Udo Hirsch); 211 Aquila (Robert Maier); 212 International Centre for Conservation Education (Patricia Bradley); 213 Bruce Coleman (Udo Hirsch); 214 Andrew Neal; 216 David Attenborough; 218 Alfred Limbrunner; 219 Kostas S. Christou; 220 *top* Aquila (Mike Mockler), *bottom* Bruce Coleman (J. L. G. Grande); 221 *top* and *bottom left* Aquila (J. Lawton Roberts; Clifford Heyes), *bottom right* Ardea (Peter Laub); 222 *left* Nature Photographers (S. C. Bisserot), *right* GeoScience Features; 222–223 *top* A. M. Duncan; 223 *bottom* Aquila (Robert Maier); 224 *top* Ardea (Werner Curth), *bottom* (both) Nature Photographers (Kevin Carlson); 225 *top* Nature Photographers (S. C. Bisserot), *bottom* (all) GeoScience Features; 226, 227 Ardea (J.-P. Ferrero); 228 *top* NHPA (Melvin Grey), *bottom* Bruce Coleman (J. L. G. Grande); 229 (both) Hans Jerrentrup.

ILLUSTRATIONS AND MAPS: *Endpapers* Ann Savage; 11 Richard Bonson; 17 Ann Savage; 28 Sarah Fox-Davies; 30 Ann Savage; 31 Sarah Fox-Davies; 35 Ann Savage; 58 Clyde Pearson; 61 Ann Savage (*after Beltran*); 68 Ann Savage (*after Mellaart*); 70 Peter Morter (*after Mellaart*); 101 Roger Kent; 124 Ann Savage; 154 Richard Bonson; 159, 169 Ann Savage; 172 Richard Bonson; 178, 207 Ann Savage.

BIBLIOGRAPHY

Part One

Beltran, A. Rock Art of the Spanish Levant. Cambridge University Press 1982

Brangham, A. N. The Naturalist's Riviera. Phoenix House 1962

Hsü, K. The Mediterranean Was a Desert. Princeton University Press 1983

Margalef, R. (ed.) Western Mediterranean. Key Environments Series. Pergamon Press 1985

Polunin, O. Flowers of Europe. A Field Guide. Oxford University Press 1969

Polunin, O. and Walters, M. A Guide to the Vegetation of Britain and Europe. Oxford University Press 1985

Part Two

Baines, J. and Malek, J. Atlas of Ancient Egypt. Phaidon 1980

Cunliffe, B. Rome and Her Empire. Bodley Head 1978

Klingender, F. Animals in Art and Thought to the End of the Middle Ages. Routledge & Kegan Paul 1971

Lauer, J-P. Saqqara. Thames and Hudson 1967

Meiggs, R. Trees and Timber in the Mediterranean World. Oxford University Press 1983

Mellaart, J. Çatal Huyuk. Thames and Hudson 1967

Seiterle, G. Artemis – Die Grosse Gottin von Ephesos. in Antike Welt no 10. Raggi Verlag Feldmeilen Switzerland 1979

Trump, D. The Prehistory of the Mediterranean. Allen Lane 1980

Part Three

Bradford, E. Mediterranean. Hodder and Stoughton 1971

Badeau, J. et al. The Genius of Arab Civilisation. Phaidon 1978

Erbstosser, M. The Crusades. David and Charles 1978

Lewis, B. (ed.) The World of Islam. Thames and Hudson 1976

Russell, W. Man, Nature and History. Aldus Books 1967

Thirgood, J. V. Man and the Mediterranean Forest. Academic Press 1981

Thomas, K. Man and the Natural World. Allen Lane 1983

Part Four

Ordish, G. The Great Wine Blight. Dent 1972

Mallinson, J. The Shadow of Extinction. Europe's Threatened Mammals. Macmillan 1978

Por, F. D. Lessepsian Migration. Springer Verlag 1985

Wirth, H. (ed.) Nature Reserves in Europe. Jupiter Books 1981

INDEX

Page references in *italic* indicate an illustration, or an illustration and text, reference. Where reference is made to a continuous sequence of pages, illustrations are not individually indicated.

Adrianople 125
Aestivation 44
Agave 186
Alaric 125
Alexander the Great *130*, 142
Algae 12, 20, *21*
al-Hambra see Granada
Allah see Islam
Alps 13, 33
America, plant introductions from 182–6
Ammanius Marcellinus 122
Ammit *84*
Amphisbaena, in Medieval myth *145*
Anchovy 20, *23*
Animal slaughter, as Roman spectacle 112–16
Annual 39
Ant 44
Antelope 13, 31
Antioch, siege of 150, *151*
Ant-lion *222*
Anubis *84–5*
Apasus 105
Aphids, and plant disease see Phylloxera
Apis bull 72, *74*, *75*, 90
 see also Bull worship
Apple, thorn 184
Arbutus 43
Aristotle 208
Artemis 105–8
 numerous 'breasts' of 106, *107*
Artemision 106
 see also Artemis
Artichoke, Jerusalem *183*
 introduction of 185
Asphodel 39, *40*
Aswan 12
Aswan High Dam 203–5
Athens, deforestation 117
 and pollution 207
Atlas Mountains *32*, 33, 52, 217
Attar of roses 157
Aubergine, introduction of 137

Baba 79
Baboon, in Egyptian mythology 79, 81
Barbarians 122, *123*, *124*, 125, 130
Barbarossa 170
Barbary 'ape' see Macaque
Barbary corsairs 169–71
Barbary fig see Opuntia
Barley 68
Barnacle goose see Goose
Bat, Egyptian fruit 52, *56*, *57*
Bean, haricot, introduction of 185
Bear 163, 208, 217
 brown *227*
 hunting of 112
 in Medieval myth 144, *147*
Beaver 163
 in Medieval myth 144
Bee, as symbol of fertility 105, *106*
 in Medieval myth 144
Bee-eater *224*
Beech 217
 in ship-building 166, 171
Beni Hasan 83
Berbers 134, 135
Bestiaries 142, 144, *145*
Birch 34
Birds, prehistoric paintings of 66
 see also Migration
Bison 163
 prehistoric paintings of 64
Black Death see Plague
Blackcap, slaughter of 198
Boar 217, *226*
 hunting of 112
 prehistoric paintings of 64
Book of the Dead *84–7*
Bracken 33
Bristleworms 19
Broom 42
Bubasti 87
Bubonic Plague see Plague
Bucephalus *130*
Bulb 39, 40
Bull 64, 217, 230
 domestication of 71
 prehistoric paintings of *61*, 64, *65*, 66, *67*
 effigies of *68*, 69, *70*
 in Egyptian painting 89
Bull capture 71, *73*, *102*, 103
Bull leaping *102*, 103
Bull worship, in Çatal Huyuk *68*, 69, *70*
 on Crete 98–104
 on Cyprus 98, *99*
 in Egypt 72–8
 in Ephesus 106, 108, *109*
 on Malta 99
 on Sardinia 99
Bull-fighting 72, 118
Burnet, thorny *42*
Cacti, introduction of 186
Camargue 36, *195*, 217, *222–4*
Camel 132
Campion 39
Cañadas 163, 165
Canary Islands 186
Cannes *195*
Caravan routes see Trade routes
Caraway *183*
Carrot, introduction of 137
Carthage 125
Cat, domestication of 79
 in Egyptian myth 79, 83
Catacombs see Saqqara, Hermopolis, Kom Ombo, Beni Hasan
Çatal Huyuk 68–71
Catchfly *38*
Cattle 58, 72
 in the Camargue *223*
 in Egyptian painting *88*, 89, *93*
 prehistoric paintings of 64
 see also Bull
Cave paintings see Painting, prehistoric
Cedar *2*, *32*, 33, 34, 217
Century Plant see Agave
Chaeronia 128
Chameleon *50*
Chariots 127
Chilli peppers, introduction of 185
Christianity 108, 111, 134, 148
 see also Crusades
Chrysanthemum *39*
Cicada 44, *45*
Clematis 229
Cockatrice, in Medieval myth *147*
Colosseum 112, 113
Columbus, Christopher 182
Constantinople, conquest by Muhammed 134
 sack of 168
 see also Crusades
Continental drift 12–15, *17*
Coral 97, 177, 180
Cordoba 135, *136*
Corm 40
Corsica 10
 extinct fauna of 31
Coto Doñana 217, *220–1*
Cotton 182
 in Islamic agriculture 157
Crab 19
 colonisation from Red Sea 181
 in Egyptian painting 89

Crete 95, 99–103, 166
endemic flora of 215
extinct fauna of 31
Crocodile 79
hunting of 112
in Egyptian myth 83
Crocus 39
Crusades 148–57, *158–9*
Cucumber, in Islamic agriculture 157
Currents see Mediterranean Sea
Cyclades 10, 95
Cyclamen 39, *40*
Cypress 44, 95
Cyprus 95, 159, 166, 171
endemic flora of 215
extinct fauna of 31

Darius, and Suez Canal 178
Dead Sea 10, *11*
Deer 29, 60, 226
dwarf 31
hunting of 112
prehistoric paintings of 64, 66, *67*
de Lesseps, Ferdinand 178, 179
Desiccation, animal defences against 44–52
plant defences against 39–44
Diana, temple of 108, *109*
see also Artemis
Dog, domestication of 71
in Mithraic imagery *110*, 111
prehistoric paintings of 65, 66
Dolphin 22, *26*
Domestication 74, 127
of cattle 69
of cereals 68
of dogs 71
of goats 71
of grape vine 97
of horse 126–9, *131*
of olive 95
of pigeon 138
of plants 68
of sheep 71
Dormouse, giant 31

Eagle, in Medieval myth 144, *145*
Imperial 217, *220*
migration 34
Earthquakes 14, 100
Earthworm 44
Eden, Garden of 7

Egypt, early settlement 72
mythology 72–94
El Djem 113, *115*
Eleonora's falcon see Falcon
Elephant 13, *28*, 29, 33, 58
dwarf *28*, 29, 31, 215
Elk 163
Elm, in ship-building 166
Embalming see Mummification
Endemic fauna 208–13, 215–7
flora 214–5
Ephesus 105–9, 118, *119*
Eucalypts, introduction of 186, 188
Evaporites see Salt deposits

Falcon, in Egyptian myth 80, 82
Eleonora's 212, *218*
see also Falconry
Falconry 138, *139*
Fertility rites 64, 66, 74, 100, 105–8
Fig, in Islamic agriculture 157
Fir 33, 95, 217
in ship-building 166, 171
Fish *24–5*
from Red Sea 180, 181
prehistoric paintings of 64
Fisheries, Mediterranean 181
decline of 199–205
Fishing *97*, 177, 200–3
Flamingo, greater *37*, 217
migration 34, 36
Flax 163
in Egyptian painting *93*
Fleas, and plague 160
Flies 44
Flint tools 58, 59, 60
Flying fish *24*
Forest 33, 60, *169*, 217
cedar 33, 34
Forest clearance, by Greeks and Romans 117, 118
during the Crusades 158
in Spain 163–5
in Italy 165
by Venetians 166, 168, 171, 173
and decline of Mediterranean ship-building 173
Forest fires 198
Fox 48
effigies of 69
in Medieval myth 144, *147*
Frankincense 132
Frog, tree *222*
Fruit bat see Bat, Egyptian fruit

Galley *167*, 170–1, *172*, 173
Gardens 136, *137*, 183
Garrigue 42, 173
Gazelle, hunting of in Egypt 89
Gecko, Moorish *50*
Genet 52, *55*
Ghar Dalam 29
Gibraltar Straits *18*, 19, 135
formation of 15
Giraffe, hunting of in Egypt 89
Glacier see Ice Age
Gladiators *113*, 114
Gladiolus *40*
Glomar Challenger 10, 13
Goat *174*
and soil erosion 173
domestication of 71
Goose, in Egyptian painting *88*, 89
barnacle, in medieval myth *145*
greylag 217
red-breasted, in Egyptian painting 89
Gorse 42
Grain 68
Granada *137*
Grape vine 96, 229
domestication of 97
in Islamic agriculture 157
see also Phylloxera
Great Bitter Lakes 177
Great Pestilence see Plague
Greece, early settlement 95
Greenfly 44, 189
Griffon 142, *146*
Guadalquivir River 208, 217
Gull, Audouin's *212*, 218
Gum cistus 43
Guyot *16–17*
see also Volcano

Hama 156, 157
Hapy 74, 90
Hare 48
Hawk, migration 34
Hawthorn 33
Heather 42
Hedgehog 29, 48
giant 31
in Medieval myth *146*
Helleborine, false *225*
Henbane 184
Herbals 140, *141*
Hermopolis 82, 83
Herodotus 178, 208

Heron, in Egyptian painting *91*
squacco *221*
Hibernation 44
Hippopotamus 29, 33, 79
dwarf *28*, 30, 31, 215
hunting of 89, 112, 113
Holly 33
Honey see Bee
Hoopoe 217, *220*
in Egyptian painting 89
Hops 229
Hornet, in Medieval myth 144
Horse 13, 58, 60, 217, *223*
Arab 132, *133*, 134
Przewalski's *126*
Tarpan 126
and fall of Rome 122–5
domestication of 126–9, *131*
in Crusades *149*, *150*, *151*
prehistoric paintings of 64, 66, *67*
Horus 80, *82*, *84*, 90
Huns 122, *124*, 125
Hunting, prehistoric 58, *59*, 60, *61*, *63*, 64, *66–7*
in Egypt 89, *90*
by Romans 112, 113
of migrant birds 198
Hyaena, hunting of 112
in Egyptian painting 89

Ibex 60, 163, 208
prehistoric paintings of 64
decline of *210*
Ibis, in Medieval myth 144
bald 210, 212
sacred 80, *81*, 82, 210
Ice Age 33–4, 60, 64
Ichkeul, Lake 217
Introductions, plant, from New World 182–6
from Australia 186, *188*
Io 105
Iris 39, *40*
yellow flag *229*
Irrigation 136, 156
Islam, ascendancy of 132–9
influence on European customs 157, 158, 160
war with Venice 168–72
see also Crusades
Islands, Mediterranean, extinct fauna of 29–31
land bridges between 33
Ivy 33

Jackal, domestication of 71
Jellyfish 22
Jerusalem 148
siege of 151, 152, *153*
Jihad 134, 148
Juniper 33, 34, 44

Kayster, river 105
Khnum *87*
Kingfisher, in Egyptian painting 89
Knights Hospitallers see Knights of St John
Knights of St John 152, 157, 159, 168, 171, 214
Knights Templar 159
Knossos 99, 100, *101*, *102*
labyrinths 105
Kom Ombo 83
Koran 132
Krak des Chevaliers 152, *154–5*, 157, 159

Ladybird 44
Larch, in ship-building 166
Lascaux 64, *65*, 66, *67*
Laurel 42
Lavender 42, *43*
Leda 105
Lemon, introduction of 137
Leopard, absence of from islands 29
effigies of 69
hunting of 112
Lepanto *169*, *170*, 171
Leptis Magna *113*, *115*, *116*, 117, 118
Lily 40, 217
martagon *225*
Lion 33, 208
absence of from islands 29, 30
hunting of 112, 113
Litter *197*, 198
see also Pollution
Lizard 48, 50, *51*, 217
agama *51*
ocellated *51*
Lobster 19, 200
Lupin *38*
Lynx 163, 208, 217, 226
decline of *209*

Macaque *2*, 33, 48, 52, *54*, 217
Mackerel 20
Magpie, azure-winged 217, *221*
Majorca 217, *228*
endemic flora of 215
extinct fauna of 31, 215
Malta 99, 171
extinct fauna of 29, 215
Maltese fungus *214*
Mammoth, prehistoric paintings of 64
Man, origin of 58
colonisation by 58–60
Neanderthal 58, 59
Mandrake 140, *141*, 184
Maquis 42, 173
Marseilles, and pollution 207
Mecca 132, 135, *136*
Mediterranean Sea, colonisation of 19–22, 179–81
currents in *18*, 19
deeps 19
evaporation from 13
formation of 10–15
salinity 19, 179
satellite photography of *17*, *18*
tides 19
Mediterranean islands see Islands, Mediterranean
Melon, in Islamic agriculture 157
Memphis 72, 77
Menna, tomb of *88*, 89–90, *91*
Mercury poisoning 206
see also Pollution
Midwife toad see Toad, midwife
Mignonette 39
Migration 20, 22
bird 34–6, 198, *199*
Minoans see Crete
Minorca, extinct fauna of 31
Minos, King 100
Minotaur *104*, 105
Mithraism *110*, 111
Mithras 111
Molluscs 19, 97
colonisation from Red Sea 181
Mongoose, Egyptian 52
Monkey see Macaque
Moors 135–7
see also Barbary Corsairs
Mosaics, Roman *97*, *112*, *115*
Mosquito 198
Moth, Jersey tiger 44, *46–7*, 48
Mouflon 71
Mouse 29, 48
in Medieval myth 144

Muhammad 132–4, 148
Mullet, red 19
grey *25*
Mummification 74–83, *85*
Muslim see Islam
Myceneans 105
Myotragus *30*, 215
Myrtle 42

Napoleon, and Suez Canal 178
Narcissus 39
Narwhal 142
Naxos 95–6
Neanderthal man see Man
Nestos River, wildlife of *229*
Nicho II, and Suez Canal 178
Nicotiana, see Tobacco
Nightingale, slaughter of *199*
Nightshade, deadly 184
woody 184
Nile 12, *176*, 177
flooding of 74, 90, 204, *205*
see also Aswan High Dam
Nuthatch, Corsican 212

Oak 95
evergreen 33
cork 43, 44
holm 44
in ship-building 166
Octopus 19, 97, 200
Oil 206
see also Pollution
Oil, olive see Olive oil
Oil, aromatic 42
Oleander 42
Olive 33, 44, 118
domestication of 95
in Greek painting *94*
Olive oil, uses of 95, *96*
as tribute 99–100, *101*
Olympic Games 128
Opuntia, introduction of 186
Orange, introduction of *137*
in Islamic agriculture *156*, 157
Orchid 39, 217
Anatolian *41*
butterfly *41*
lizard *225*
Provence *41*
yellow Bee *41*
Oreganum 42
Orontes, river 156, 157
Oryx, hunting of in Egypt 89
Osiris *85*
Osiris-Apis, Temple of 77
see also Apis bull
Osprey 217
Ostrich, hunting of in Egypt 89
Otter 217
Owl, eagle 226

Painting, prehistoric *59*, 60, *61*, *63*, 64–8
Egyptian *88*, 89, *90–3*
Greek *94*, *96*, *104*
Roman *113*
Paprika 185
Paradise 134, 136, 148
see also Gardens
Pardel see Lynx
Parkinson, John 183
Parpallo 59
Pasht 79, *87*
Peacock 138
Pearls 97
Peloponnese see Myceneans
Pelican, in Medieval myth 144
Dalmatian, decline of *210*
white, decline of 210
Peony 33, 217
Atlas *53*
Peppers, introduction of 184
Pesticides 206
see also Pollution
Petaloudes Valley *46*
Philip II of Macedonia 128
Phocaea 208
Phoenix, in Medieval myth 142, *146*
Phylloxera 189–93
Pigeon, domestication of 138
Pillars of Hercules see Gibraltar Straits
Pimpernel *39*
Pine 33, 44
Maritime 33
stone 217
in ship-building 171
Pink 39
Piperi see Sporades, Northern
Pipits, slaughter of 198
Plague 160–3
Plaice 19
Plankton 20, *21*
Plant diseases 188, 189
see also Phylloxera
Plato 117
Pliny 117, 140
Plitvice Lakes National Park 217, *225–7*
Plough *120*
in Egyptian painting *93*
Pollution 203–6, *207*, 230
Pomegranate, in Islamic agriculture 157
Posidonia 20
destruction of 203, *204*
Potato *183*
introduction of 184
Potato, sweet 182, *183*
Pottery, Cypriot 98, *99*
Pratincole *224*
Prickly pear see Opuntia
Przewalski's Horse see Horse
Ptah 72
Temple of 74, 77
Pyrenees 33

Rabbit 48, 60, 217
Ram, effigies of 69
Rat, black 160, *161*
brown 160
see also Plague
Raven 33
in Mithraic imagery *110*, 111
Red Sea *176*, 177
marine colonisation from 179–81
Reindeer 58
Reptiles 48, *49–51*
Rhinoceros 33
prehistoric paintings of 64
Rhodos 159, 168
Jersey tiger moth on see Moth, Jersey tiger
Rhône 12
Robin 33
slaughter of 198
Rock-rose 42
Roman empire 111–19, *124*
decline and fall of 122–5
Rosemary 42

Sacrifice, animal 69, 108, 111
Sage *42*
Sahara 33
Salamander 140, *143*, *146*
Salinity see Mediterranean Sea

Salt deposits 10, *11*, 12, 13
Salt lakes 10, *11*
Samos *8*
Santorini 103
Saqqara *73*, 77, 80–3
Sardine 20
  decline of 205
Sardinia 10
  extinct fauna of 31
Scarab beetle 72
Scorpion, in Mithraic imagery *110*
Sea-squill *40*
Seal, monk 22, *27*, 218, *219*
  decline of 208, 210
Seeds, as defence against
    desiccation 38, 39
Sekhmet *86*
Sen-nedjem, tomb of *92–3*
Serapeum 77, *78*
Serapis, (Osiris-Apis) see Apis bull
Seth 79
Sewage 203, 206
  see also Pollution
Shark 19
  blue *25*
Sheep, domestication of 71
  Merino 163–5
Shrew 29
  in Egyptian mythology 82
Shrines, in Çatal Huyuk *68*, 69, *70*
Shrubs 39, 42
Sicily, extinct fauna of 31
  over-fishing 201
Skate 19
Skink 50, 217
  ocellated *51*
Snail 44, *45*
Snake 48, 79
  in Medieval myth 144
  in Mithraic imagery *110*, 111
  leopard *49*
  Montpellier 48, *49*
  sand boa *49*
  viperine 215–16
Sobek 79, *86*
Soil erosion 118, 165, 173
Solanaceae (family) 184
Sousse, mosaics *97*
Spices 132, 157, 173, 182
Spinach, introduction of 137
Sponge 97
Spoonbill *221*
Sporades, Northern 217, 218
Spruce 217
  in ship-building 166
Squid 19
St Paul, expulsion from
    Ephesus 108
Stilt, black-winged *228*
Stock 39
Stockholm Conference 206
Stork, white *36*
  migration 34, *35*
Strabo 178
Stromboli *14*, *15*
Suez Canal *176*
  building of 178–9, *180–1*
  passage of marine organisms
    through 179–81
Sugar cane, in Islamic
    agriculture 157
Sunbathing, hazards of 196
Swallow, migration 34
Swan, giant 31
Swordfish 20

Tamarisk 214
Tarpan see Horse
Tawaret *87*
Thebes 88
Thermopylae 118
Theseus *104*, 105
Thoth 80, 82, *83*, *84*
Thrushes, slaughter of 198
Thyme 42
Thysdrus see El Djem
Tidal waves 100
Tides see Mediterranean Sea
Timsah, Lake *176*, 177
Tit 33
Toad, midwife, Majorcan 215, *216*
Toadflax *38*
Tobacco, introduction of 185
Tomato, introduction of 185
Tortoise 29, 48
  giant 31
  spur-thighed *49*
Tourism 194–7, *207*
  and litter *197*, 198
  effects on wildlife 198
  see also Pollution
Trade routes 105, 132, 166–8,
    169, 173
Tuber 41
Tuna 20, 200, *201*
Tunny see Tuna
Turks, war with Christendom 168,
    170–1
Turtle, leatherback 22
  loggerhead 22, *26*

Unicorn 142
United Nations Environment
    Programme 206
Urban II, Pope 148

Valltorta *61*, *63*
Vandals *124*, 125
Venice, hegemony of 166–73
Vetch *38*, 39
Vine see Grape vine
Visigoths *124*, 125
Volcano 14, *16–17*
Vulcan 14
Vulture, black 217, *228*
  effigies of 69

Walnut 33
  in ship-building 166
Warbler, migration 34
  Cyprus 212
Wasp 44
  in Medieval myth 144
Weasel, effigies of 69
  in Medieval myth 144, 146
Whale, killer 22
  pilot 22
  sperm 22
Wheat 68, 118
Wheatear, black-eared *224*
Willow, dwarf 34
Wine see Grape vine
Wolf 208, 226
  domestication of 71
Woodpecker 33
  black 226
Wrack 20

Zeus 105

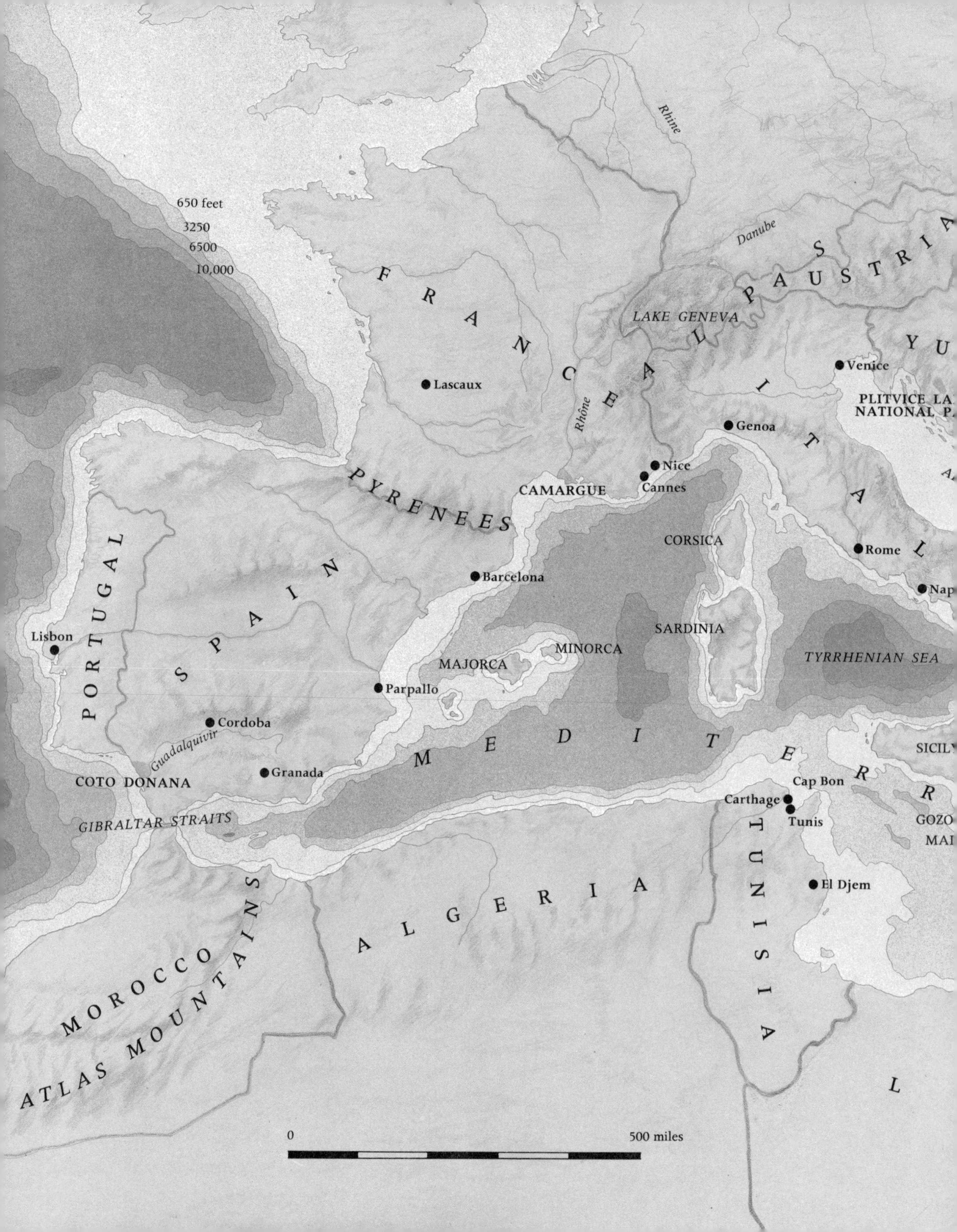
650 feet
3250
6500
10,000
Rhine
Danube
FRANCE
ALPS
AUSTRIA
LAKE GENEVA
ITALY
Venice
PLITVICE LA
NATIONAL P
Lascaux
Rhône
Genoa
Nice
Cannes
CAMARGUE
PYRENEES
CORSICA
Rome
Nap
Barcelona
PORTUGAL
SPAIN
Lisbon
SARDINIA
MINORCA
MAJORCA
TYRRHENIAN SEA
Parpallo
Cordoba
Guadalquivir
Granada
COTO DONANA
MEDITERR
SICIL
Cap Bon
Carthage
Tunis
GOZO
MAI
GIBRALTAR STRAITS
TUNISIA
El Djem
ALGERIA
MOROCCO
ATLAS MOUNTAINS
0
500 miles